Introduction to
PLANT
DISEASES

Identification and Management

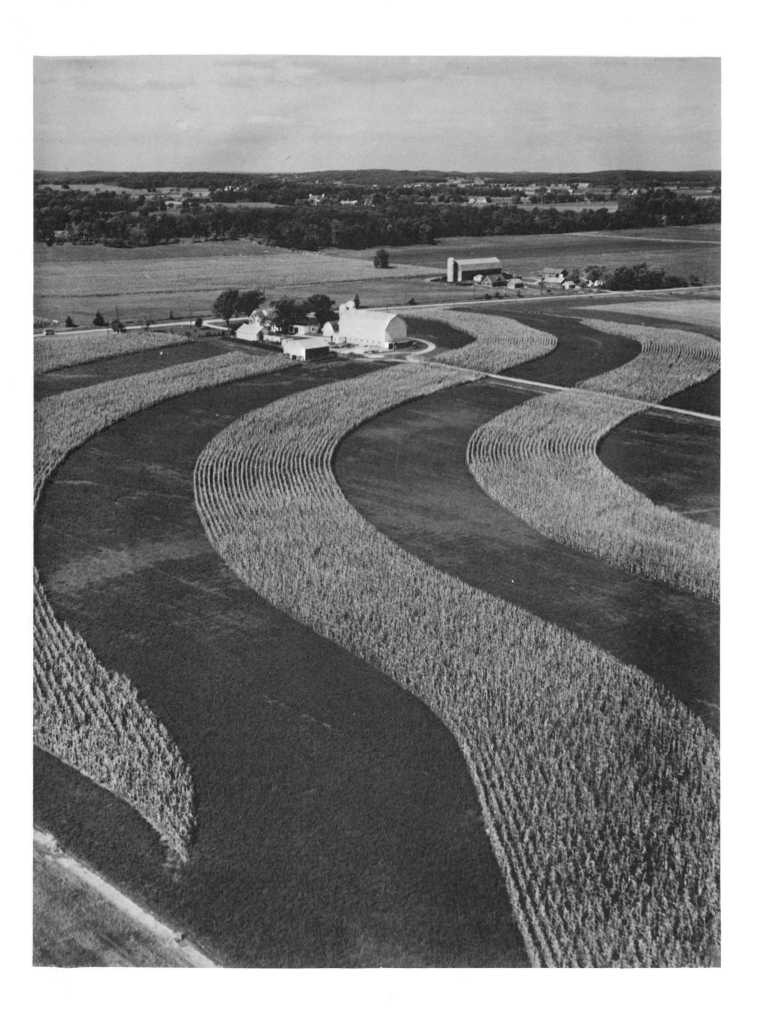

Introduction to
PLANT
DISEASES

Identification and Management

G. B. LUCAS
C. L. CAMPBELL
L. T. LUCAS

Department of Plant Pathology
North Carolina State University
Raleigh, North Carolina

An **avi** Book
Published by Van Nostrand Reinhold
New York

Frontispiece: A contoured, strip farm in the midwestern United States. Contouring and strip cropping help prevent soil erosion and decrease the spread of soilborne pathogens.

Cover design by Lou Sardella.

An AVI Book
(AVI is an imprint of Van Nostrand Reinhold.)

Library of Congress Catalog Card Number 85-1325
ISBN 0-87055-473-5

Printed in the United States of America

Van Nostrand Reinhold
115 Fifth Avenue
New York, New York 10003

Van Nostrand Reinhold International Company Limited
11 New Fetter Lane
London EC4P 4EE, England

Van Nostrand Reinhold
480 La Trobe Street
Melbourne, Victoria 3000, Australia

Macmillan of Canada
Division of Canada Publishing Corporation
164 Commander Boulevard
Agincourt, Ontario M1S 3C7, Canada

16 15 14 13 12 11 10 9 8 7 6 5 4 3 2

Library of Congress Cataging-in-Publication Data

Lucas, G. B.
    Introduction to plant diseases.

    Bibliography: p.
    Includes index.
    1. Plant diseases.   2. Micro-organisms, Phytopathogenic
—Control.   I. Campbell, C. L. (C. Lee)   II. Lucas, L. T.
III. Title.
SB731.L83   1985   632′.3   85–1325
ISBN 0–87055–473–5

*This book is dedicated to
the proposition that we should
make people aware of the importance of
plant diseases and how they affect
our lives daily.*

Contents

Contents (continued)

Contents (continued)

Contents (continued)

Foreword

The growing technology associated with agricultural production, whether it be that of large-scale commercial farms or of the homeowner and amateur gardener, requires textbooks and other sources of information that present the fundamentals of a complicated subject in understandable terms. One such complicated subject is plant pathology, a knowledge of which is requisite for profitable food production.

The responsibility for writing a textbook in plant pathology usually rests upon those who are willing to devote time and energy beyond the normal requirements of their assigned duties. Such an effort is seldom profitable and also has certain professional and emotional hazards. Criticism is inevitable.

It is therefore gratifying for me to endorse this text, which can be used to introduce the beginner and amateur agriculturalist to the fundamentals of plant pathology. The material has been thoughtfully put together and reflects years of experience by the authors in teaching undergraduate students, extension agents, and growers.

Agricultural consultants, chemical company representatives, and sales personnel will find this book useful in educating their clientele about plant diseases. It should also meet a special need in technical programs in biological sciences and agriculture in high schools, technical colleges and introductory courses at 4-year colleges.

One of our departmental objectives has been to develop courses for the general public on plant diseases. A major problem has been the lack of an organized body of information suitable for this clientele. This book should help in this endeavor and enable more people to appreciate the nature and importance of plant diseases.

Dr. Robert Aycock

Department of Plant Pathology
North Carolina State University
Raleigh, North Carolina

Preface

We wrote this elementary text to fill a need for a book on plant diseases for students who do not have a background in the biological sciences. Thus, we designed this book so that it can be used in high schools, technical institutes, or colleges; we believe it will also be useful to agricultural teachers, farmers, home gardeners, landscape architects, and pesticide salesmen.

Our book provides practical information that helps in the successful management of diseases on food, fiber, and landscape plants. The main emphasis is on four aspects of plant pathology: (1) how to recognize that a plant has a disease; (2) how to determine the cause of the disease; (3) how to manage diseases; and (4) where to seek additional help with disease problems. Our purpose is to give examples of specific diseases and their management to illustrate general principles that can be applied to a wide range of problems, rather than give an exhaustive review of all important diseases.

The list of suggested readings at the end of the book provides the reader access to source material on plant pathology and its importance in our society. These references may be used as a basis for further study of plant diseases and their management.

We are indebted to our colleagues and students for criticisms, concepts, discussions, encouragement, and ideas throughout the writing of this book. We are especially grateful to Dr. Robert Aycock, the members of the Department of Plant Pathology, and Dr. H. B. Craig, Director of the Agricultural Institute at North Carolina State University. We wish to thank the many members of the Agricultural Extension Service, Agricultural Research Service, and Academic Affairs of the School of Agriculture and Life Sciences at North Carolina State University whose material we used. We especially wish to thank Charles R. Harper, W. Marvin Williams, Michael R. Newnam, Eileen Sutker, Alan S. Windham and Judy P. Sinn who assisted in the preparation of the manuscript, and Polly Carlsen and Joyce Denmark for typing the manuscript.

Finally, we are obliged to our wives—Vernelle, Karen, and Joy—whose patience, encouragement, and love sustain us and give life meaning.

G. B. Lucas
C. Lee Campbell
L. T. Lucas

Related AVI Books

BREEDING FIELD CROPS, 2nd Edition
  Poehlman
BREEDING VEGETABLE CROPS
  Bassett
COFFEE: BOTANY, BIOCHEMISTRY AND PRODUCTION OF BEANS
AND BEVERAGE
  Clifford and Willson
FUNDAMENTALS OF ENTOMOLOGY AND PLANT PATHOLOGY, 2nd
Edition
  Pyenson
FUNDAMENTALS OF LANDSCAPING AND SITE PLANNING
  Root
HORTICULTURAL REVIEWS, VOLS. 1–7
  Janick
INTRODUCTION TO FRESHWATER VEGETATION
  Riemer
LANDSCAPE PLANTS IN DESIGN: A PHOTOGRAPHIC GUIDE
  Martin
PLANT BREEDING REVIEWS, VOLS. 1–3
  Janick
PROTECTING FARMLANDS
  Steiner and Theilacker
SITE ENGINEERING FOR LANDSCAPE ARCHITECTS
  Strom and Nathan
STATISTICAL METHODS FOR FOOD AND AGRICULTURE
  Bender, Douglass and Kramer
TISSUE CULTURE OF TREES
  Dodds
VEGETABLE GROWING HANDBOOK, 2nd Edition
  Splittstoesser
VEGETABLES IN THE TROPICS
  Tindall
WORLD VEGETABLES
  Yamaguchi

Agriculture, Plant Diseases, and Human History

Hendrick Van Loon, the famous historian, once wrote, "The history of mankind is the story of a hungry creature in search of food." This observation has been true since man first appeared on the earth thousands of years ago and is true today. Recorded history is a chronicle of the suffering, death, and upheavals in society following disease, pestilence and famine. At any given time, there is enough food to feed all the people on earth for only about three weeks. Even that small amount of food is unequally distributed and famine stalks the world. One-third of the earth's population awakes hungry each morning and goes to bed hungry at night. Hunger is a way of life in many developing countries. As stated in a 1980 report to President Carter: "For hundreds of millions of the desperately poor, the outlook for food and other necessities of life will be no better by the year 2000. For many it will be worse. . . . Hunger and disease will claim more babies and young children and more of those surviving will be mentally handicapped by childhood malnutrition." This dismal and alarming forecast underscores two of the most important problems facing mankind in the twentieth and twenty-first centuries—overpopulation and hunger.

What has all this to do with plant diseases? Well, man depends upon plants for his very existence. Only green plants and some blue-green algae can convert solar energy into food. Therefore, man and other animals exist on earth as guests of the plant kingdom. Three-fourths of the total world food supply is drawn from the grasses, and all human civilization rests on the cultivation of cereal grains. Most of the human food supply worldwide is derived from the following ten crops:

- Wheat
- Rice
- Potatoes
- Maize
- Sweet potatoes
- Cassava
- Millet, sorghum
- Soybean
- Pulses (beans, peas)
- Sugar cane

In addition to the food we eat, plants provide food for the animals we use for meat, for work, and for pleasure. Lumber from the forests provides shelter and furniture; plant fibers furnish clothing, fabrics, rope, and paper. Other plants supply man with medicines. Many beverages and drugs—including beer, cocoa, coffee, whiskey, wine, quinine, opium, and tobacco—are derived from plants. All fossil fuels, such as oil and coal, on which our current high standard of living depends, originated from plants or from animals that lived on plant substances. Plants beautify our surroundings, purify the air and water, stabilize soil, prevent floods and erosion, and protect our natural resources. Flowers, fields, and forests help satisfy our longings for beauty, fulfill our aesthetic cravings, and provide relaxation and recreation.

Plants and animals coexist. Plants provide food for animals, either directly or indirectly, and in turn use the carbon dioxide released into the atmosphere during the respiration of animals, burning of fuels, and processes of decay. Man has learned to cultivate certain plant species for use as food, shelter, or clothing. These relatively few species have been selected from wild plants and improved to produce larger yields of superior quality. As a result of improvements in farm management, pest control, herbicides, and irrigation and fertilization methods since 1950, U.S. farmers have produced food in abundance for our people year after year, with some left over for export.

However, plants do get sick. They have diseases. Plant diseases cause famine and famine results in human suffering and diseases. People who are undernourished and weak from hunger are easy prey to cholera, pneumonia, stomach disorders, and other infectious diseases and parasites. Famine and epidemics ride side by side. People huddled together in unsanitary, crowded cities and ghettos are easy prey to unrest and panic. As Seneca said, "A hungry people listens not to reason nor cares for justice, nor is bent by any prayers."

Throughout human history, reductions in crop yields from diseases, pests, or bad weather have had severe effects on the whole human race and on specific, local populations. Despite the improved plant cultivars (cv) and more efficient cultural techniques that now are available to grow crops, many people in the world still do not have an adequate diet, and hunger is their constant companion. Among the many plant diseases that have influenced world history are the following:

- Bacterial wilt
- Chestnut blight
- Coffee rust
- Downy mildew of grape
- Ergot of rye
- Potato late blight
- Rice blast
- Root knot
- Stem rust of wheat
- Tobacco mosaic

The Study of Plant Diseases

Plant pathologists study the causes and development of plant diseases and try to develop practical management methods. The study of plant diseases is known as *plant pathology* or *phytopathology*. These words are derived from the Greek words *phyto,* meaning plant, *pathos,* meaning disease, and *logos,* meaning to study.

By understanding the causes of plant diseases, learning to recognize them, and developing ways to reduce crop losses, plant pathologists help to increase food and fiber production and to protect our food supply. In this way, they combat hunger, reduce suffering, and make the world a better, safer place for all humans.

Plant pathologists are usually concerned with disease in many plants rather than with disease of a single plant. Occasionally, diseases on a single plant such as a large shade tree or ornamental shrub will receive special attention, whereas the loss of a single plant in a large wheat field usually would not be noticed. A medical doctor, in contrast, usually deals with disease on an individual basis.

The first step in a disease management program is to identify the disease. It is essential to correctly diagnose the problem before selecting and using a pest management practice. Sometimes it is easy to isolate a disease organism but it may take years to identify the real cause of the trouble. Some farmers have had enough experience with plant pests and are familiar enough with disease symptoms that they can correctly diagnose plant troubles. However, when some unknown problem occurs, the grower should seek professional help.

Many different areas of science and agriculture are involved in the study of plant diseases. Plant pathology began as an offshoot of *botany*, the study of plants. To understand diseases and develop management methods, the physiology, anatomy, taxonomy, and growth of plants must be studied. The following sciences also contribute to the understanding of the causes of plant diseases:

- *Bacteriology*—the study of bacteria
- *Biochemistry*—the chemistry of plant and animal life
- *Biotechnology*—the collection of industrial processes that involve the use of biological systems, including in some cases genetically engineered microorganisms
- *Chemistry*—the study of the composition, properties, and structure of substances
- *Edaphology*—the study of soils
- *Engineering*—the study of machines and engines
- *Entomology*—the study of insects
- *Genetics and Plant Breeding*—the study of sexual reproduction in plants
- *Meteorology*—the study of weather
- *Mycology*—the study of fungi
- *Nematology*—the study of nematodes
- *Physics*—the study of matter and energy and their interaction
- *Statistics*—the description of data using mathematics
- *Virology*—the study of viruses
- *Weed Science*—the study of weeds and their control.

Plant pathologists use centrifuges, complicated chemical equipment, computers, electron microscopes, greenhouses, light microscopes, photography, spectrophotometers, and sterile growth chambers to gain information leading to the development of more efficient plant-disease management methods.

Plant Disease Losses

Although there are at least 50,000 diseases of economic plants and new diseases are discovered every year, it is difficult to accurately assess losses from disease. However, it is safe to say that annual losses in the United States average about 15% of the total agricultural production, or more than $15 billion. In North Carolina alone losses are estimated to be worth more than $300 million per year. Disease causes another $30 million in economic loss in terms of the replacement value and increased maintenance costs of ornamentals and turf in that state. Similar losses occur in other states.

Plant diseases vary widely in the damage they cause. *Annihilating* diseases may wipe out a crop completely; for example, chestnut blight annihilated the American chestnut throughout most of its range in the United States. *Devastating* diseases, such as southern corn leaf blight, may be severe for a few years then gradually subside. *Limiting* diseases, including many root diseases, build up to the point where it is not economical to grow a crop in a particular field. Then there are *debilitating* diseases that occur almost every year and weaken crops, resulting in depressed yields and poor quality. No crop ever produces its full yield. Many bad things can happen to a crop between sowing the seed and using the harvest.

Plant diseases do their damage in different ways. One major effect of disease in plants is reduced plant vigor resulting in decreased yields. Another major impact is reduced quality; for example, wheat rust infection reduces the size of grain kernels and lowers carbohydrate content; diseased apples are discolored and unsightly; decayed wood is weak or unusable. Some plant pathogens contaminate or infest soil to such an extent that a desirable field cannot be used for growing a principal crop unless the soil is fumigated or rotations established.

Disease management expenses constitute one of the major losses associated with plant disease. Some diseases, such as apple scab and brown patch of turf, can be kept in check only by repeated applications of pesticides. This requires time, equipment, and labor, all of which add to the cost of production. Misuse or misapplication of pesticides by careless or ignorant people results in environmental pollution, injury, and/or death to many living things, including humans. The cost of pesticide misuse is difficult to measure.

Some plant pathogens form *toxins* in stored grain, peanuts, hay, or other products. If such moldy products are fed to animals or eaten by humans, they may become sick or die. These costs are difficult to estimate accurately.

Additional information on the effects of plant disease on world history is given in Chapter 3 and in the discussions of individual diseases.

Effects of Modern Agriculture on Plant Diseases

The great plains of the United States and Canada are known as the breadbasket of the world, and elsewhere throughout the United States farmers produce a wide array of crops upon which our society rests. The happy combination of suitable soil and climate in large areas of the country, skilled growers and farmers, and researchers and educators dedicated to obtaining and disseminating new knowledge has led to very high levels of agricultural production. However, the war against plant diseases never ends. The primary mission of agricultural research centers is to develop production practices and new disease-resistant cultivars that reduce the adverse effects of pests and unfavorable environmental conditions on plant growth.

Despite the development and widespread use of high-yielding cultivars of improved quality and of new plant protection technologies, the possibility of severe damage from disease and pests has not been eliminated. Indeed, some farm practices actually increase the severity of plant disease. Poor land management, sanitation, and agronomic practices may favor the buildup and survival of plant pathogens or predispose crop plants to increased damage from disease.

Failure to destroy crop residues or weed hosts that may harbor disease agents, improper land preparation and erosion prevention, use of infested seed, improper use of fertilizers and irrigation, growing the same cultivar in the same field year after year (*monoculture*), failure to control weeds and insects—all these may contribute to disease increase. Introducing a highly susceptible crop into a region occupied by an aggressive pathogen, or permitting a virulent pathogen to invade an area where a susceptible crop is grown, may result in epidemics and huge crop losses. Plant breeders may develop a cultivar, resistant to a major disease, that turns out to be susceptible to a minor disease; the minor disease may soon become a major one.

Failure to maintain *genetic diversity* in crop plants also can exacerbate disease problems. For example, when a desirable high-yielding, disease-resistant cultivar or several closely related cultivars resistant to only one strain of a pathogen are introduced over a wide area, new strains of the pathogen may arise in these "pure" stands of genetically similar plants. An epidemic occurs, huge losses result, and after a few years the "new" desirable cultivar is no longer usable—the risk of disease is too great.

What Is a Plant Disease?

A plant disease may be defined as any disturbance that prevents the normal development of a plant and reduces its economic or aesthetic value. A disease interferes with the normal function of some part of the plant, resulting in lower yields or reduced quality. For the farmer, this leads to less income; for the consumer, reduced food supplies of lower quality and higher prices; and for the homeowner and turf manager, lower aesthetic value, less beauty and higher maintenance costs.

Diseases are caused by living agents (microorganisms and parasitic plants) and by nonliving agents (environmental factors, faulty nutrition, and chemical substances). The microorganisms and parasitic plants are called *pathogens* or *biotic agents*. The nonliving environmental and nutritional factors and chemical substances are called *abiotic agents*. In some diseases both pathogens and abiotic agents are involved.

A plant disease develops over a period of time, whereas an injury that may be caused by an insect feeding on a plant or damage from equipment or machinery occurs immediately or over a very short period of time. Usually a disease has been developing for several days before any symptoms on the plant are visible.

Recognizing Plant Diseases

The visible reactions of a plant to disease are called *symptoms.* These include wilting, stunting, yellowing, death, and abnormal growth of part or all of a plant. A diseased plant is recognized by comparing it to a healthy plant. However, although plants rarely are free of all diseases during their life span, they may appear healthy. Therefore, to determine if a plant has a disease, we must know as much as possible about the growth, development, culture, and appearance of normal, healthy plants. Agronomy, horticulture, and forestry are sciences that deal with the growth and culture of plants.

Accurate diagnosis of a plant disease depends on the identification of the pathogen or causal agent. Microscopic examination or isolation of the pathogen is often necessary for diseases caused by microorganisms. The identification of the causal agent of a disease that has not been described previously or one that is not well known must meet four criteria, called Koch's Postulates:

1. The pathogen or abiotic agent must be consistently associated with the disease.
2. The pathogen must be isolated in pure culture or an abiotic agent identified.
3. When the pathogen or abiotic agent is applied to healthy plants, the same symptoms develop as on the diseased plants originally examined.
4. For diseases caused by biotic agents, the same pathogen must be isolated from the diseased plants that were inoculated with the isolated pathogen (step 3).

Useful Strategies for the War Against Plant Pests

The war against plant pests, especially plant diseases, is similar in many respects to war between nations. The same basic principle applies: to wage war successfully and contain the foe requires a sound strategy. However, the first thing to realize in planning strategy is that we will never win the war against plant pests. The war is never-ending. But even if we cannot win the war, we can still win battles. Sound strategy requires knowledge and judgment, because the more we know about our enemy the more chances we have of defeating him. A strong, capable intelligence agency (research) is needed to gather information and knowledge about plant pests, their capabilities and resources, about crop plants and their defensive capability, and about the battlefield (environment) where the encounters occur. The object is to provide crop plants with every possible advantage in the attack-defense equation, to contain the foe at the least possible cost, and to minimize losses.

To wage war costs money. Therefore, growers must know how to expend their resources wisely. If they do not, costs and losses may be intolerably excessive. Battles may be lost and then the money spent for disease management is wasted. Every dollar spent in pest management should return $3–$4 to growers.

Continuous pest management is basic to efficient production. Plant pests are extremely resourceful and constantly challenge the judgment, skills, and resources of the defense. Hence, growers must know not only the current capabilities of the foe at any given time and place, but also their own potential to mount future offensives. They must make suitable preparations in advance for battles that are bound to come. Sometimes the foe can launch a surprise attack without warning; the defense breaks down and epidemics occur.

Strategy in the war against plant diseases involves two main objectives: to reduce the initial amount of the pathogen and to reduce its rate of increase and spread. Tactics involves the development of suitable weapon systems by which strategic objectives can be achieved. Logistics involves the marshalling of weapons and resources so that they are available in adequate amounts at the right time and place. Trained personnel are essential. Execution involves the safe, efficient, and effective operation of management practices. Growers need to know what each weapon (management practice) is for, how and when it is to be employed, and how effective it is. They should use chemicals wisely and avoid polluting the environment.

Overall strategy and tactics may be complex, especially where more than one plant pest is involved. Successful management may require the use of various combinations of weapons that assist and strengthen each other just as military battles involve certain combinations of artillery, air power, and tanks in support of the infantry.

In war the defense sometimes breaks down. Casualties occur and heavy losses are sustained. This requires a reevaluation of the present strategy and the development and employment of new weapons designed to prevent recurrent disasters. The war against plant diseases must not be abandoned if a hungry world is to be fed.

2

Causes of Plant Diseases

Infectious Plant Diseases

Infectious plant diseases are caused by living (*biotic*) agents, or pathogens. These pathogens can be spread from an infected plant or plant debris to a healthy plant. *Microorganisms* that cause plant diseases include *nematodes, fungi, bacteria,* and *mycoplasmas.* We also classify *viruses* and *viroids* as biotic agents since they must have living cells for reproduction and are composed of nucleic acid and protein. Some higher plants that produce seeds are parasitic on other plants and are considered to be pathogens. Each type of plant pathogen is discussed briefly in this chapter. More detailed information is given in the chapters dealing with the individual diseases. The relative sizes of various types of pathogens are illustrated in Fig. 2.1.

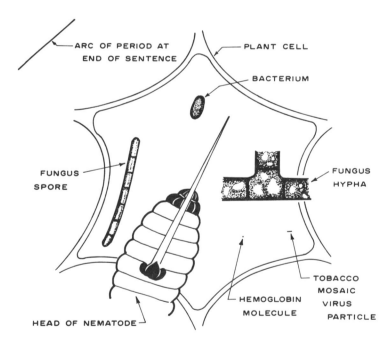

Fig. 2.1. **Relative sizes of plant pathogens. The actual sizes (average) are as follows: diameter of period at end of sentence = 1/80 in. = 0.3 mm = 300 μm (micrometers); diameter of plant cell = 15 μm; diameter of nematode head = 4 μm; diameter of fungus hypha = 2 μm; length of bacterium = 1 μm; length of tobacco mosaic virus particle = 0.3 μm = 300 nm (nanometers); diameter of hemoglobin molecule = 5 nm.**
Adapted from G. N. Agrios

Nematodes

Nematodes are small eel-like worms, most of which are too small to be seen with the unaided eye. Nematodes feed on plants by means of a miniature hypodermic-like structure called a *stylet,* which is used to suck liquid nutrients out of plant cells. Nematodes reproduce by laying eggs that hatch into young nematodes called *larvae* or juveniles. The larvae develop through a series of four molts, in which the outer skin is shed, until they become adults. Some nematodes can complete their life cycle in less than 30 days.

The most important nematode pathogens injure plant roots, but different types of nematodes can feed on almost any part of a plant. Some nematodes feed mainly on the outside of plant tissues (*ectoparasites*), whereas others wriggle inside the tissues and feed from within (*endoparasites*). Nematodes cause damage to plants by injuring cells, removing cell contents, or changing normal plant growth processes. Chemicals that are used to kill nematodes are called *nematicides.*

Fungi

Fungi are small threadlike organisms composed of tiny filaments called *hyphae.* Individual hyphae are composed of strands of simple cells so small they cannot be seen without a microscope. The tangled mass of intertwined hyphae is collectively called *mycelium.* Fungi may develop into colonies such as those often seen on bread or vegetables that have been stored improperly or for a long time. Some fungi grow inside plant tissues or organic debris and cannot be seen unless magnified; others develop into large structures such as mushrooms or toadstools.

Most fungi produce microscopic "seedlike" spores, which are often spread from one plant to another by wind, water, or man. The spores land on a suitable substrate (plant) and germinate. The hyphae may penetrate the plant surface directly, through natural openings or through wounds. Fungi damage plant tissues by producing toxins, enzymes, or growth-regulating substances that alter or destroy plant tissues. Many different types of chemicals, called *fungicides,* have been developed to kill fungi that cause plant diseases.

Bacteria

Bacteria are small, single-celled organisms that possess rigid cell walls. Bacterial cells must be magnified about 400 times to be seen through a light microscope. Bacteria reproduce by a process called binary *fission* in which one cell divides into two cells. Under suitable conditions, with plenty of available nutrients, some bacteria can divide about every 30 min. If every bacterial cell divided into two cells every 30 min, a total of 8,388,608 cells would be produced after 12 hr from a single cell.

The rapid multiplication of bacteria and the production of toxins and enzymes that destroy plant tissues contribute to the damage caused by bacteria. Only rod-shaped bacteria cause plant diseases. They enter plants through wounds or natural openings. Some bacteria produce growth-regulating substances that cause the surrounding tissues to grow abnormally. Antibiotics called *bactericides* have been used to manage some plant diseases caused by bacteria.

Mycoplasmas

Mycoplasmas, first shown to cause plant diseases in the 1960s, are small single-celled organisms that occur inside plant cells. Mycoplasmas do not have rigid cell walls; each is enclosed by a triple-layer cell membrane. These cell membranes can be easily ruptured if the cells are not in a favorable environment.

Many diseases that were earlier thought to be caused by viruses, and were called *yellows* diseases, are caused by mycoplasmas. These organisms are spread from one plant to another by a *vector,* usually an insect such as a leafhopper. Mycoplasmas can also be spread in infected plant parts. Tetracyclines are antibiotic substances that have been used to kill some mycoplasmas in plants.

Viruses and Viroids

Viruses are very small particles, usually less than 300 nm in length, composed of nucleic acid and protein. They can be seen only under an electron microscope at magnifications of 2000–3000 times. Viruses that infect plants have several different shapes: some are spherical; some are, comparatively speaking, long rigid rods; some are long flexuous rods; and some are short thick rods. Nucleic acid (usually ribonucleic acid, or RNA) composes the center of the particles and is surrounded by and connected to a protein coat. The RNA in viruses directs certain enzyme systems in plant cells to manufacture components used to produce more virus particles.

Viruses cause diseases in plants by diverting energy and structural components, normally used in plant growth, into reproductive processes for the virus. Viruses are spread from one plant to another by insect or nematode vectors or by man himself. Another major means of spread is in infected plant parts. Viroids are smaller than viruses. They act like viruses but the infectious particle is simply a strand of RNA and contains no protein, only RNA. Therapeutic chemicals that inactivate plant viruses and viroids are not yet available.

Parasitic Plants

Parasitic plants are higher plants that reproduce by seed. Most of these parasitic plants have modified rootlike structures that attach to plant tissues to obtain nutrients and water, but do not have root systems that can absorb nutrients from the soil. Parasitic plants weaken the host by using nutrients that would normally be utilized by the host plant. Mistletoes are some of the most frequently seen parasitic plants. In some forest trees attacked by mistletoes, the plant tissues may be so disorganized that the wood is weakened and the branches deformed.

Disease Complexes

Frequently plant diseases, especially those induced by soilborne pathogens, involve more than one pathogen. When this double trouble happens, increased damage to or death of the plant may occur. A plant may be able to resist the attack of one parasite, but when two or more attack the same plant it may be overwhelmed. Whenever the combined damage of two or more pathogens acting together is much greater than the sum of the damage caused by the pathogens acting alone, we call this a *plant-disease interaction*. For example, tomatoes suffering from root knot may be killed more rapidly if attacked by the Granville-wilt bacterium. The combination of cold injury, bacterial canker, improper pruning, and root-rotting organisms causes short life of peach trees. Air pollutants predispose ponderosa pine to attack by bark beetles.

Environmental extremes—such as too much or too little water, temperature or relative humidity (RH)—often affect disease development; nutritional excesses or deficiencies may increase the susceptibility of plants to some diseases. Insects, nematodes, fungi, weeds, and even man often transport pathogens from one area to another. Satisfactory management programs for disease complexes require accurate diagnosis of the problem and the combined use of various practices if the affected crop is to be grown profitably.

Noninfectious Plant Diseases

Noninfectious plant diseases are caused by nonliving (*abiotic*) agents that cannot be transmitted from one plant to another. These agents are usually environmental, nutritional, or chemical factors. Environmental extremes in temperature, moisture, or light often are unfavorable for plant growth. Nutritional deficiencies and excesses cause disease-like problems that must be correctly identified if they are to be prevented or eliminated. Chemically induced plant diseases result from improper soil pH, improper use of fertilizers and pesticides, chemical spills, and air pollution. In recent years air pollutants from automobiles and industrial processes have been shown to drastically affect some plant species and the quality of water in many lakes (this phenomenon is called *acid rain*).

Names of Pathogens and Diseases

All living things have common names and scientific names. The common names may vary from country to country and from one language to another. However, in 1753 the Swedish naturalist, Linneaus, proposed the binomial system of nomenclature in which each type (species) of organism is given a Latin name consisting of two parts. Latin was chosen because it was a widely known language used by scholars in many countries. The first part of the binomial is the *genus* or generic name, and the second is the *species* or specific name. Also, the name, or an abbreviation of the name, of the person who first described the plant or animal is often written following the binomial.

Thus, the name for the white rose is *Rosa alba* L. and signifies that Linneaus originally gave the name *Rosa alba* to this particular kind of rose. Species names are underlined in typed material and generally italicized in printed material, as they are in this book.

Once a Latin binomial has been applied to a certain animal or plant, the same name can never legitimately be used to apply to any other species. The scientific name of one of the most widespread root-knot nematodes is *Meloidogyne incognita* (Kofoid and White) Chitwood. Kofoid and White first described this nematode, but Chitwood later revised the genus and gave it a new name that had priority. Fungi and bacteria also have Latin binomials. However, the naming of viruses is still being discussed.

Oftentimes the scientific names of plants and animals have many syllables and are difficult to pronounce. Despite this disadvantage, the binomial system has proved useful and efficient for more than 200 years.

Pathogens are given common names, such as root-knot nematodes, to use in nonscientific settings in place of the scientific name, i.e., *Meloidogyne incognita.* Diseases are usually given a common name that describes some part of the symptoms or sometimes a portion of the pathogen name is used. Brown patch is a common name descriptive of the brown dead areas in turfgrasses that are caused by the fungus *Rhizoctonia solani* Kühn. Root knot refers to the galls on roots that are caused by the root-knot nematode. An example of a pathogen name being used as part of the common name is Fusarium wilt, a disease of many kinds of plants that is caused by the fungus *Fusarium oxysporum* (Schlecht) Snyder and Hansen.

Common names for viruses usually indicate the host affected or symptoms produced by that virus. For example, cucumber mosaic causes a mosaic symptom on cucumber and some other plants. See Chapter 17 (Naming of Viruses section) for a fuller discussion.

3

History of Plant Pathology

In Ancient Times

Plant diseases have been recognized for centuries. The ancient Romans worshipped Robigus, the god of wheat, and sacrificed red dogs to him at the Robigalia (a great festival) to save the grain from the destructive red rust. Aristotle, Plato's student, recorded plant diseases as early as 350 B.C., and his colleague Theophrastus observed and speculated about diseases of cereals, legumes, and trees. Evidently, plant diseases were destructive in ancient times and the people lived in fear of famine. Thus, the belief in supernatural causes of plant diseases is not surprising. The idea of Divine Punishment for man's sins, as evidenced by the occurrence of plant disease and the resulting human suffering, is very old. This belief was encouraged by religious leaders who, because of their own ignorance, held similar beliefs and it also helped preserve the tradition of the infallibility of the Church. Man blamed diseases on evil spirits, displeasure of the gods, or unfavorable positions of the stars or moons.

Early biblical writings (for example, the books of Amos, Deuteronomy, and Kings I in the Old Testament) tell of mildews, blasts, and blights. In Revelation, the last book of the New Testament, St. John the Divine writes of the Four Horsemen of the Apocalypse—War, Famine, Disease, and Death—who ride the earth and contest man's dominion over it. Shakespeare in *King Lear* tells of the foul fiend who walks at night and mildews the white wheat. In the Middle Ages, outbreaks of ergot fungus in Europe poisoned the bread made from infected grain. Convulsions seized those who ate the rotten bread; they went mad and died painful deaths. These unfortunate victims of a plant disease were considered possessed and, at times, accused of engaging in witchcraft.

For 2000 years after the time of Theophrastus, little was added to the knowledge about plant diseases. Their nature and causes were shrouded in superstition and speculation.

From the time of Christ until the Renaissance little was written about plant diseases. The Middle Ages that brought European civilization to a near standstill also enveloped the study of botany and plant pathology. However, increased trade between Europe and the Near East led to the development and rise around 1000 A.D. of a rich middle class of merchants and tradesmen with leisure time available. Many were educated and became doctors, lawyers, or teachers. In their spare time, some became amateur naturalists and studied all forms of life. And they wrote down what they saw. Down through the centuries since the time of Christ there were a few men in every generation who wanted to find out more about their world. However, throughout the Western world these curious naturalists were hampered by the Christian church, which sought to regulate scientific inquiry, demanded a literal interpretation of the Bible, and preserved the doctrine of the infallibility of church teachings.

Gradually through the centuries the belief that filth, decay, and disease produced living entities spontaneously was accepted. This gave rise to the theory of *spontaneous generation.* For example, it was generally accepted that fungi were the result of plant diseases, not the cause. Similarly, rats were thought to be formed in moldy rags or garbage left in corners.

Until the Middle Ages and well into the Renaissance, most people could not read nor write. The few books available were copied by hand and communications were slow and expensive. New knowledge about plant diseases, or anything else, traveled slowly and to relatively few people. About 1450, Johann Gutenberg invented the printing press, an accomplishment that ranks with the invention of the wheel or the harnessing of fire in the ascent of man. With this remarkable machine, multiple copies of books could be made and distributed widely. Education of the masses had begun.

From 1600 to 1800

Two centuries after Gutenberg, in about 1683, Anton von Leeuwenhoek, a Dutch lens grinder, built a crude microscope that enabled him to produce and see magnified images too small to be visible to the naked eye. Leeuwenhoek examined many things: raindrops, saliva, and rotting substances. What he saw delighted and surprised him. He discovered protozoa, bacteria, and other microorganisms. Leeuwenhoek's microscope opened up a whole new world to observation and propelled the science of microbiology forward.

In 1728, Duhamel de Monceau in France described the saffron disease of crocus. He conclusively showed that the disease was caused by the fungus *Rhizoctonia,* and that the fungus was contagious—it could spread among plants and cause epidemics. He also proposed sound management measures for the disease. Unfortunately, his work received almost no attention from other scientists and was lost to plant pathology and the rest of the scientific world. Duhamel was ahead of his time. One can only wonder how much further advanced medical and botanical sciences would be if Duhamel's accurate statement of the germ theory had been widely publicized and accepted in 1728.

In 1743, J. T. Needham first described plant-parasitic nematodes in wheat galls, but a hundred years passed before M. J. Berkeley, in England, observed the root-knot nematode.

P. A. Micheli, an Italian, in 1729 used the microscope to study many fungi. He showed that "seeds" of fungi grew and produced more "seeds." In 1755, Mathieu Tillet, in France, proved experimentally that wheat bunt is contagious and could be prevented by seed treatments. Slowly but surely careful observers were showing that plant diseases did not arise spontaneously but were caused by living things—"germs."

In 1807, Isaac Benedict Prevost from Switzerland proved conclusively that bunt, a disease of wheat, is caused by a fungus and could be controlled by dipping seed in copper sulfate. Prevost pointed out the importance of the environment in the development of the disease. He was also ahead of his time. His great contribution was rejected by the authorities, but his seed treatment became widely used despite lack of approval by the so-called "learned men."

The Golden Age of Biology, 1840–1900

A mysterious and terrible *potato murrain* appeared in western Europe and the United States from 1830 to 1850. In 1845, this potato murrain, or late blight, as it is known today, destroyed the Irish potato crop. As a result, a million people died of starvation and malnutrition and a similar number of Irish citizens emigrated from Ireland to the United States, mostly to the northeastern states. The consequences of this tragedy forced scientists to investigate the disease and find ways to control it. Finally, Heinrich Anton DeBary (Fig. 3.1), one of the outstanding biologists of his time, proved experimentally that a fungus was the cause of the late blight of potato and had his work accepted by the general scientific community. At the age of twenty-two he published a book presenting overwhelming evidence that fungi are the causes not the results of plant diseases. DeBary's book demolished the theory of spontaneous generation and helped establish the *germ theory of disease,* which states that germs cause disease. This was only one of DeBary's many triumphs. DeBary is called the Father of Plant Pathology not only because of his scientific ability, but also because he was a great researcher, compiler, and

Fig. 3.1. Heinrich Anton DeBary (1831–1888). A medical doctor and famous German mycologist who showed that fungi cause plant diseases, DeBary is known as the Father of Modern Plant Pathology.

educator. He trained more than 60 scientists from many countries who became prominent in the field of plant pathology and mycology.

In the Golden Age of Biology from 1840 to 1900, the foundations of modern biology were laid and much progress was made. Julius Gotthelf Kühn, a German, published the first plant pathology text in 1858. Oscar Brefeld, Robert Koch, Louis Pasteur, and Dr. R. J. Petri (one of Koch's assistants) introduced techniques for growing organisms in pure culture and helped to establish the germ theory of disease. Another German, Robert Hartig, devoted his life to the study of forest tree diseases and published two books on the subject. Hartig is widely known as the Father of Forest Pathology. Thomas Jonathan Burrill, at the University of Illinois, showed in 1878 that fire blight of apple and pear was caused by a bacterium.

Fig. 3.2. Erwin Frink Smith (1854–1927). Through his work at USDA, Smith established much of the basic knowledge on plant diseases caused by bacteria, earning him the title Dean of American Plant Pathologists.

In 1882, Frenchman Pierre-Marie-Alexis Millardet developed Bordeaux mixture for the control of downy mildew of grapes. Erwin F. Smith began his long and fruitful career with the U.S. Department of Agriculture (USDA) in 1886 and became one of this country's greatest bacteriologists and plant pathologists (Fig. 3.2). Also in the 1880s, Adolph Mayer in Holland demonstrated that tobacco mosaic could be spread by juice from infected plants. Dimitri Iosifovich Ivanovski, a Russian, in 1892 claimed that the entity that caused mosaic was smaller than a bacterium. Several years later, Martinus Willem Beijerinck working in Holland, demonstrated that the infectious substance that caused mosaic increased in tobacco plants. He was the first to use the word *virus,* from the latin word for poison, in relation to a plant disease. In 1884, Hatsuzo Hashimoto, a Japanese rice grower, recorded that a leafhopper transmitted the agent (he did not know it was a virus) causing the dwarf disease of rice.

The Twentieth Century

In the twentieth century, plant pathology matured as a science with startling and rapid discoveries. Such men as E. C. Stakman, L. R. Jones, and H. H. Whetzel attracted and trained many students in the United States; pioneer researchers, including Erwin F. Smith, forged ahead in new areas of plant pathology. Jones in the 1920s restated Prevost's idea that environment was important in the development of plant diseases; this time the idea was accepted. N. A. Cobb, G. Steiner, B. G. Chitwood, and their co-workers expanded nematology.

In 1935, W. M. Stanley won the Nobel Prize for demonstrating that tobacco mosaic virus was a chemical (a protein) that could reproduce itself in living cells. F. C. Bawden and N. W. Pirie showed in 1936 that Stanley's protein, in a more purified liquid crystalline state, was a nucleoprotein, and that a virus was made up of a nucleic acid, such as ribonucleic acid (RNA), and protein together. G. A. Kausche and his colleagues in 1939 first saw virus particles in the amazing electron microscope.

In the early 1940s, scientists formulated the first carbamate fungicides; chemical nematicides came into use in the 1950s. Systemic fungicides were developed and first used in the 1960s and the *sterol-inhibitor* fungicides were introduced in the 1970s and early 1980s.

In 1963, J. E. Vanderplank published his book on epidemics and management of plant diseases. His ideas permitted scientists to describe mathematically the factors that contribute to plant disease development.

In 1967, Y. Doi and co-workers in Japan observed a new kind of organism (mycoplasmas) in infected plants; these were transmitted by leafhoppers. In 1971, T. O. Diener discovered viroids, which are smaller than viruses and thus the smallest known infectious agents of plant disease.

In addition to the development of new pesticides and the description of newly found pathogens in the 1960s and 1970s, several new areas of research began to contribute to the understanding of plant diseases. Scientists today are studying the biochemistry of compounds and enzymes involved in the development of plant disease. Other scientists are working on the epidemiology of plant diseases. Through these studies, they are seeking a more complete understanding of pathogen spread and disease development in plant populations over large geographic areas. In addition, recent advances in biotechnology have enabled plant pathologists to utilize additional sources of disease resistance and to shorten the time it takes to develop disease-resistant cultivars. Based on advances in these areas, scientists will be able to provide more efficient management measures for plant diseases.

Today, the flood of published research continues to increase. The storage, retrieval, and use of this information is one of the most important problems in science. New discoveries in computer science, biotechnology, genetics, physiology, and many other sciences, all are having their impact on plant pathology.

A hungry world awaits progress in the war against plant diseases—a war that will never end.

4

Development of Plant Diseases

The Plant Disease Triangle

For a plant disease to occur, a pathogen, a susceptible host plant, and a favorable environment must be present over a certain period of time. This is known as the *disease triangle.* As shown in Fig. 4.1, each side of the triangle represents one of these factors and the amount of disease is indicated by the height. The effect of reducing the amount of the pathogen by various management practices also is illustrated in Fig. 4.1: the amount of disease is less even though a susceptible host and a favorable environment are present. Environmental factors can also be adjusted to reduce the amount of disease that develops: for example, the temperature and RH can be altered in greenhouses and the pH or nutrient status adjusted by modifying the soil. On the other hand, if a resistant host grows in a favorable environment in the presence of a pathogen, disease will not develop because the host is not susceptible and the triangle is not complete. The disease triangle concept can be used to visualize many disease management approaches.

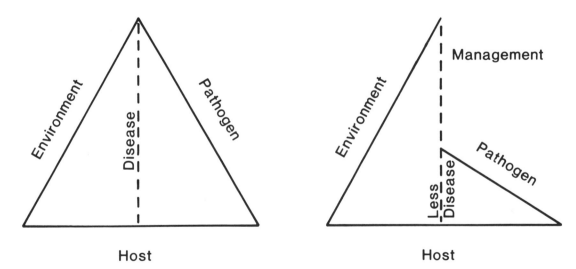

Fig. 4.1. *Left:* The plant disease triangle, including the three factors necessary for disease to appear. *Right:* The effect of management in reducing disease.

Since each side of the disease triangle can occur in a favorable or unfavorable state, there are eight possible combinations of pathogen, host, and environment. Only one of these combinations will lead to disease. For example, the combination of a susceptible host and the pathogen without a favorable environment will not result in disease nor will the combination of a favorable environment and aggressive pathogen without the presence of a susceptible host. Can you list the other five combinations that will not result in disease? Only the combination of a susceptible plant, a favorable environment, and a pathogen will result in disease. Therefore, mathematically, the chance that disease will develop is 1 in 8.

This concept helps to explain why most plants are not affected by the many different diseases that exist. It is obvious that most plants are not affected by the many thousands of diseases that occur because one of the factors in the disease triangle is absent. For example, spores from many different fungi on nearby crops might land on the leaves of an oak tree at the edge of a field, but very few, if any, of these spores cause disease on the oak tree because it is resistant to these fungi: no susceptible host is present, thus no disease develops.

The Disease Cycle

The term *disease cycle* is used to describe the relationship of a pathogen to its host in an environment and the development of a disease over time. It involves the survival of the pathogen during periods that are not favorable for disease development, dispersal of the pathogen to susceptible hosts, growth of the pathogen on or in plants, development of disease symptoms, secondary spread of pathogens from diseased plants to healthy plants and, to complete the cycle, pathogen survival. The study of the survival and spread of pathogens and the development of a disease in a population of plants over time is called *epidemiology*.

Pathogen Survival

Many pathogens can survive in the absence of a susceptible host during conditions unfavorable for disease development. In cold climates pathogens of annual plants may survive from one growing season to the next in plant debris or in seeds or in the soil. In frost-free regions, others survive on growing plants. Many pathogens of perennial plants survive in infected tissues of the host plants or in propagating materials. When conditions are favorable, these pathogens begin to grow and spread to susceptible plants or tissues. Pathogens such as viruses or some rusts can survive in *alternate hosts*.

The survival mechanisms just described apply to pathogens that are reduced to a low level if susceptible plants are not grown in the area for several years. Other pathogens can survive for many years in the soil in the absence of a susceptible host. These pathogens may be able to compete with saprophytic organisms in the soil or produce resistant structures that do not germinate unless susceptible plants are present. It is obvious that successful management methods for these different types of pathogens will vary.

Pathogen Dispersal

Some pathogens have the ability to spread only a few centimeters, whereas others can be transported for many kilometers. Pathogens are usually spread by wind, water, soil, in plant parts, or by a vector.

By Wind

Tiny, fragile fungal spores are well adapted to air travel and are blown near and far by the wind. Spores of rust fungi have been detected at altitudes of 10,000 m and are blown for at least 1500 km on turbulent air currents. But spores of other pathogens may be carried only a few centimeters by wet wind or spattering rain.

Winds strong enough to blow dust particles have enough power to carry insects, nematode eggs, and billions of invisible fungal spores and bacterial cells capable of causing disease outbreaks to distant places. Some fungi possess specialized fruiting structures that propel the spores into air currents, thus assuring wider distribution.

By Water

Rainwater transports soil and debris infested with fungi and bacteria along rows, terraces, and ditches and from infested fields into disease-free fields, seed beds, drainage ponds, creeks or streams. Infested water used for irrigation or transplanting may carry spores to new locations and cause disease outbreaks. Some fungi and bacteria are spread in blowing rain. These types of pathogens must remain moist until they land on a susceptible plant or they will die. Other pathogens are spread in water as propagating units such as *zoospores* (swimming spores) or in debris that may be washed from place to place. This group includes soil-inhabiting pathogens that can survive in the soil and includes nematodes, fungi, bacteria, and parasitic plants.

On Plant Material

Many pathogens are spread by infected transplants or cuttings. Infected seed is responsible for many disease outbreaks, and the use of contaminated seed has been responsible for introducing plant pathogens into new agricultural areas. Man himself is the principal agent for spreading diseased plants to new areas in the course of normal production and marketing practices. During the sorting and grading of fruit, vegetables, and ornamentals for market, discarded culls, leaves, and other infected parts can be a source of pathogens to start disease outbreaks.

By Vectors

Insects, nematodes, fungi, and animals are important agents that transport pathogens from one place to another. Many of the fungi, mycoplasmas, and viruses that cause diseases are spread by insect vectors from diseased to healthy plants (Table 4.1). Some viruses and mycoplasmas reproduce in their insect vectors; this unusual relationship makes management difficult.

Table 4.1. Some Insect Vectors of Plant Pathogens

Insect	Pathogen transmitted
Aphids	Tobacco mosaic virus
	Potato virus Y
	Cucumber mosaic virus
	Tobacco vein-mottle virus
	Sugar cane mosaic
Bark beetles	Dutch elm disease fungus
	Oak-wilt fungus
	Chestnut-blight fungus
Cucumber beetles	Anthracnose fungus
	Cucurbit-wilt bacterium
Flea beetles	Corn-wilt bacterium
Honey bees, wasps, flies	Fire-blight bacterium
Leafhoppers	Beet curly-top virus
	Aster yellows mycoplasmas
Plum curculio	Peach brown-rot fungus
Seed corn maggot	Soft-rot bacterium
Thrips	Tomato spotted-wilt virus
White flies	Tobacco leaf-curl virus
	Cotton leaf-curl virus

Pathogen Growth on the Host Plant and Appearance of Symptoms

The process whereby a pathogen comes in contact with the host is called *inoculation.* For disease to develop, the pathogen must penetrate the host, invade tissue, and begin to grow and establish an intimate relationship with the host.

Pathogens penetrate host tissues in various ways. Nematodes pierce plant cells with their *stylets* and suck out the cell contents. Spores of many fungi get inside the plant by growing through natural plant openings such as *stomates* and *hydathodes;* some spores produce *infection pegs* that can penetrate the leaf surface directly. Most bacteria gain entry into the plant through natural openings or through injuries. Many viruses are injected into plants by insect vectors while feeding; others penetrate through the microscopic injuries resulting from the wounding or rubbing of plants.

Some pathogens penetrate plant surfaces but do not cause disease because an incompatible (unsuitable) host-pathogen relationship develops. A resistant plant may respond to penetration by undergoing complicated biochemical and physiological changes that kill the pathogen or prevent its spread through the plant. For a disease to develop, a relationship must be established so that a pathogen can continue to grow and colonize host tissue.

As the pathogen continues to grow in the plant, symptoms will begin to appear at a certain time after penetration, depending on the particular pathogen and host plant involved. Many biochemical and physiological changes have usually occurred in the plant before symptoms develop. Visible symptoms are the responses of the plant to disease processes that have been occurring over a period of time. The time between inoculation and symptom development is known as the *incubation period.*

Some diseases, including many of the root rots, have only one cycle during a growing season. Other diseases, particularly ones that occur in aerial parts of the host, develop secondary, or repeating, disease cycles during a growing season. Recurrent crops of spores, maturing every week or 10 days, provide inoculum to infect nearby healthy plants. Disease develops quickly and may cause severe damage over a short period of time. This rapid disease increase is called an *epidemic.*

The disease cycle is completed when the pathogen reaches the survival stage. Many pathogens produce specialized structures, such as *sclerotia* and *resistant spores,* that enable the pathogen to resist freezing, drying, or other adverse conditions.

Environmental Effects on Disease Development

For hundreds of years farmers blamed disease on bad weather; many people still think bad weather causes disease. Today we know this is not true, but we also know that weather greatly affects the frequency and rate of disease development, the amount of disease, and the damage caused by epidemics. The reaction of the host and pathogen, independently and in combination, to the many environmental factors determine if and how much disease will develop.

Temperature

Different plants and pathogens thrive at different temperatures. Thus, it is no surprise that there are warm-weather (bacterial wilt) and cool-weather (potato late blight) diseases. Many diseases develop at temperatures optimum for pathogen development but unsuitable for the host. For example, the fungus *Gibberella (Fusarium) zeae* causes a root rot of both wheat and corn. The fungus does most damage to wheat (a cool-weather crop) at warm soil temperatures, whereas it does most damage to corn (a warm-weather crop) when soil temperatures are low. Temperature also affects growth and sporulation of pathogens. Some downy mildews and rust fungi produce few spores when day temperatures rise above 30°C and remain above 20°C at night.

Moisture

The three major sources of moisture are rain, dew, and irrigation, but fog and high RH also supply moisture. The occurrence of many diseases in a particular region is correlated with the amount and distribution of rainfall. In some cases, rainfall determines not only disease severity but also whether a disease will occur at all. Many diseases of young tender tissues are favored by high moisture. The spores of many fungal pathogens will not germinate unless enveloped in a film of water, and free water in plant tissues is favorable for the growth of bacteria. On the other hand, hot, dry weather prevents many epidemics from developing. Dry soils inhibit the growth of nematodes and other soil pathogens. The amount of moisture in soil determines the amount of free oxygen, and therefore controls soil aeration.

Wind

Wind can spread pathogens and their insect vectors from field to field and over long distances. Wind has a greater effect when accompanied by rain. Windblown rain releases fungal spores and bacteria from infected tissue, blows them to new locations, and deposits them on wet plants where they may start new infections. On the other hand, dry winds may depress disease development by rapidly drying plant surfaces and decreasing the RH. Windblown dust particles may injure tender leaves, stems, and flowers. Plants rubbing against each other during wind storms may be wounded, thus making them more easily infected.

Light

The quality, intensity, and duration of light affect growth of both host and pathogen. Plants grown in reduced light are often weak and etiolated. As a result they are more susceptible to infection. Changes in day length, which profoundly affect plant growth, may decrease or increase susceptibility to various diseases. Photoperiod and light intensity also influence inoculum survival (some spores are killed by exposure to sunlight for 1 hr), abundance of sporulation, the ability of spores to penetrate the host, and length of the incubation period.

Nutrition

Nutrition affects the rate of plant growth and the ability of plants to defend themselves against disease. For example, overfertilization with nitrogen results in rapid vegetative growth, the formation of tender succulent tissues, and delayed maturity, thus increasing susceptibility to pathogens that attack rapidly growing plants. On the other hand, lack of nitrogen results in slow growth and more rapid aging which makes plants more susceptible to attack by those pathogens that attack weak, slow-growing plants.

The amounts of other major nutrients—phosphorus, potassium, and calcium—and of micronutrients available to plants also may affect their susceptibility to disease attack and resistance to disease.

Soil Type and pH

There are many types of soil, each composed of a solid, liquid, and gas phase in varying proportions. This complex environment—teeming with many different kinds of microorganisms, some beneficial and others harmful—has profound effects on both plant growth and disease development. Light sandy soils with low organic matter are especially favorable for growth of many nematode pathogens, whereas heavier soils with high organic matter are less conducive to nematode diseases. Soil structure and texture affect water-holding capacity and soil temperature. As a result, the incidence of damping-off is increased in heavy, cold water-logged soils.

The acidity or alkalinity (as measured on the *pH* scale) also affects disease development, principally of root diseases. Some soil pathogens attack plants over a wide pH range, whereas some (club root of cabbage) do most damage in "sour" soils with a low pH and others (black root rot of clover) in "sweet" soils with a high pH.

Cultural Practices

The kind of crops (row vs. broadcast), cultivar used, number of plants per hectare, methods of cultivation (no-till vs. cultivated), and harvest method (hand or machine) also can affect disease incidence. Therefore, farmers should use the most efficient agronomic practices for a given crop that do not provide conditions favorable to disease outbreaks.

Forecasting Plant Diseases

Fig. 4.2. **This battery-powered microcomputer and microenvironmental monitor identifies weather conditions favorable for development of peanut leafspot. Recommendations ("Spray" or "Do Not Spray") are displayed on the lighted panel and are based on accumulated data on temperature and RH.**
Courtesy J. E. Bailey

Since time immemorial, farmers have worried about the weather and used all means at their disposal to protect their crops by predicting future weather and disease outbreaks. Farming is a risky business, and accurate weather and disease forecasts often lead to increased profits.

Disease forecasts are useful to farmers in the practical management of disease in their crops. The ultimate purpose of forecasting is to minimize disease losses. Farmers must balance three factors in disease management: risk, cost, and benefit. For example, if forecasts indicate that the risk of disease is low, there is no need to invest in the expense of spraying with pesticides; if the forecast predicts a high disease risk, then it may be wise to start a spray program. Forecasting provides the means for determining when and where a given management practice should be applied. If a fungicide is not needed, why go to the expense of applying it?

One of the peaceful spin-offs of the space program is the use of satellites to monitor weather patterns around the world, permitting forecasters to make more accurate weather predictions, both local and regional, for the next few hours, the next week, or longer. Also, the rapid progress in computer science and equipment permits huge volumes of information to be analyzed quickly and efficiently. The small "grey boxes," placed in farmers' fields or orchards to record temperature, rainfall, light, wind, and RH data, are really small computers that help in predicting disease outbreaks (Fig. 4.2).

Forecasts Based on the Weather

The incidence of most diseases is highly dependent upon the weather, and some early forecasting methods were based solely on weather patterns. For example, late blight of potato occurs during rainy weather, with high RH, and cool temperatures (below 22°C). Conversely, there is little chance of late blight breaking out during periods of hot, dry weather. Therefore, successful late-blight forecasting can be accomplished by keeping accurate temperature and RH records at critical times during the growing season. Forecasts based on weather can indicate when diseases are and are not likely to occur.

Forecasting based on weather assumes that the pathogen is always present in a locality and disease will break out if suitable weather develops. Sometimes the pathogen is *not* present in which case the disease will not break out even under favorable conditions; this is one of the drawbacks of weather-based forecasting.

Forecasts Based on Inoculum Potential

Some forecasts are based on the amount of the pathogen available to infect the host. Spore traps are used to determine the presence and numbers of airborne spores; for soilborne pathogens, assays of soil or plant tissues determine how many propagules are present in the soil or plant litter and debris. These data are then used to calculate the chances of disease outbreaks. Forecasts based on the amount of pathogen present assume that weather favorable for disease development will occur. This assumption does *not* always hold true.

Forecasts Based on Type of Pathogen and Rate of Increase

As a result of Vanderplank's contributions to the study of epidemiology, another type of disease forecasting is available. If the type of pathogen present and its rate of increase under different weather conditions is known, the chances of disease occurring can be predicted with some accuracy based upon measurements of the amount of inoculum present in a given area.

Undoubtedly, with future advances in computer science, sophisticated equipment, mathematical modeling, and more accurate collection of weather data, disease forecasting services will be increasingly valuable to farmers and help in safeguarding production of our daily bread.

5

Integrated Pest Management

Plant disease management is just one phase of farming. It is an integral part of the whole program of good farming, but it is not necessarily the most important part—unless diseases get out of hand every year and build up to the point where farmers can make no crop at all. And this has happened! Usually disease management is just one part of a complex system that must be performed so that good enough yields can be obtained to make some money every year. Practices used to manage diseases must be integrated with the other practices necessary to produce a healthy crop. Disease management must fit in with efficient insect and weed management, rotations, irrigation, cultivation, harvesting, storage, marketing, and every other sensible practice that the farmer finds necessary to get the most profitable yields. Management of diseases is part of a system called integrated pest management (IPM).

Although farmers have used some IPM principles for years, recent emphasis has been on the economics of pest management practices. Farmers know instinctively that when they plant a crop they are gambling that they will reap a harvest (Fig. 5.1). They cannot always predict problems from diseases, insects, markets, and weather. They know that some things are beyond their control. But what farmer would not select new practices if there were a good chance of getting four dollars back for every dollar expended? Today we call this the cost/benefit ratio or cost/risk ratio. Most farmers will try new production practices, especially if there is a good risk (high statistical probability) that they will be profitable.

Evolution of Pest Management Practices

The history of man is filled with his attempts to gain increasing control over the environment. About 8000 B.C., man learned how to cultivate grain crops. It was a remarkable advance that freed him from a nomadic existence, but this subsistence agriculture was hampered by unsuitable weather, low soil fertility, and low-yielding varieties. Lack of machinery and ignorance of yield-increasing agronomic practices prevailed for hundreds of years.

Through the centuries, however, many cultural and physical practices evolved for protecting crops from plant pests through trial and error. Such measures as burning or plowing under crop refuse, rotating crops, planting in the most favorable season, using healthy seed, isolating crops, pruning, and

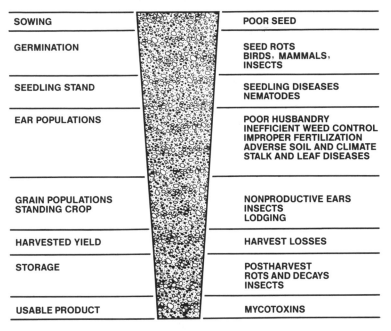

POTENTIAL YIELD
200 Bu/Acre

SOWING	POOR SEED
GERMINATION	SEED ROTS BIRDS, MAMMALS, INSECTS
SEEDLING STAND	SEEDLING DISEASES NEMATODES
EAR POPULATIONS	POOR HUSBANDRY INEFFICIENT WEED CONTROL IMPROPER FERTILIZATION ADVERSE SOIL AND CLIMATE STALK AND LEAF DISEASES
GRAIN POPULATIONS STANDING CROP	NONPRODUCTIVE EARS INSECTS LODGING
HARVESTED YIELD	HARVEST LOSSES
STORAGE	POSTHARVEST ROTS AND DECAYS INSECTS
USABLE PRODUCT	MYCOTOXINS

ACTUAL YIELD
100 Bu/Acre

Fig. 5.1. Why corn yields are low. No crop ever produces its full yield. Many bad things can happen to a crop between sowing the seed and using the harvest.

managing water and fertilizer have been used for hundreds of years to improve the chances of making a good crop. As biological knowledge grew in the eighteenth and nineteenth centuries and as pest problems became more severe as a result of more complex agriculture and the introduction of pests into new areas, the search for more effective pest management methods intensified. The discovery in 1882 that spraying with Bordeaux mixture would reduce damage from downy mildew on grapes encouraged scientists to look for other chemicals that would kill pests. Since then many highly effective agricultural chemicals have been developed and widely used. These chemicals gave farmers a new ability to reduce pest populations at a relatively low cost. The widespread adoption of these chemicals for managing pests, along with the development of disease-resistant cultivars, led to a general decline in the use and study of many basic cultural and biological tactics for reducing pest populations to a level where profitable yields are possible.

By the late 1950s, however, it became apparent that unacceptable levels of crop loss from pests were still occurring despite tremendous improvements in crop production and protection methods. In addition, a variety of problems have caused concern about various agricultural technologies and future food production. In particular, the development of pests resistant to chemicals, crop losses and stress due to the phytotoxicity of pollutants, shifts in the prevalence of pest species, and the harmful effects of some pesticides on nontarget organisms including humans have all raised questions about the efficacy and advisability of certain pest management practices.

Modern agriculture can overcome many of these problems, but its increased complexity and intensity demand unprecedented precision in the management of crop pests. Furthermore, the realization that plant diseases are natural components of agriculture that must be dealt with continuously fortifies the concept that diseases can never be "controlled" or eliminated. To win the war against plant pests requires that they be "managed" in a process that incorporates the latest knowledge into the management program as rapidly as possible. The important principles of pest management are summarized in Table 5.1.

Many farmers are increasingly aware of the need for efficient farm management. Changes in farm structure, cultural practices, the price of food, farm inputs, interest rates, the limited choice of crops, special processing and marketing requirements, governmental restraints, and increased transportation costs—all affect farmers' decisions to use a certain practice. In addition, it is obvious that no strategy to increase food production on a sustaining basis can afford to disregard environmental and ecological principles. Efficient agricultural production in the short term cannot justify eventually destroying the ecological basis on which our entire food production system rests. The aim should be agricultural production without environmental destruction.

Continuous pest management undergirds efficient and profitable crop production. Every season of the year, and during certain seasons almost daily, the farmer needs to be actively fighting pests. All farmers must be taught to manage pests, for every farmer's practices affect the community of farmers. Some pests can be held at bay only when all farmers in the community participate in a continuous pest-management program.

Table 5.1. *Important Pest Management Principles*

- Plant diseases and/or pests can never be eliminated, only managed at an economically acceptable level.
- Pest management requires knowledge and judgment: know the enemies (the pests), know the battleground (the environment), know the crop.
- Provide the crop with every possible advantage. Use all weapons in the arsenal of integrated pest management.
- Continuous pest management is basic to efficient production.
- Epidemiology is basic to development of a sound pest management strategy.
- Contain the foe at the least possible cost and minimize losses. Every dollar spent in pest management should return $3–$4. There are short-term gains and long-term gains.
- There are two main strategic objectives: reduce the initial inoculum, and reduce the rate of increase and spread of the pest.
- Use proper tactics. Develop and employ suitable weapons to carry out strategy. These include chemicals; resistant cultivars; seed treatment; crop rotation; proper planting date, seeding rate, and depth; proper harvesting and storage; plowing under of crop debris.
- Use effective logistics. Deploy weapons and resources at the proper time, in the correct amount, and the correct way. Trained personnel are essential.
- Beware of pollution. Use chemicals wisely.

Revival of the IPM Concept

The phrase *pest management* was first used in the late 1950s to describe the combined use of biological and chemical methods of insect control. It recognizes that pest species are a normal part of the environment and that (1) most crops can tolerate or compensate for certain levels of infection without incurring economic loss, (2) many pests play beneficial roles in the environment, and (3) eradication of pest species is seldom possible.

The term *integrated* was added later to indicate the need to use all relevant biological, chemical, and cultural tactics in the management of insects, plant pathogens, weeds, or any pest that lessens productive yields. Integrated pest management is a comprehensive and systematized approach that includes crop protection in the total cropping systems based on input from many disciplines.

Although IPM can be defined in numerous ways that describe the concept from different viewpoints, the following definition was adopted by the North Carolina State University Committee for Extension Agent Training in IPM: "Integrated Pest Management is the selection, combination and use of pest management actions on the basis of predicted consequences." It considers the economic, social, and political consequences of pest problems and seeks to utilize knowledge and techniques from all crop production and protection disciplines in the management of pests. Two important components of IPM are (1) an information-gathering system for *monitoring* insect, pathogen, and weed populations, and (2) the *planning* and *implementation* of a systematic crop management program. Integrated pest management seeks to contain pest populations at levels that permit farmers to stay in business. It uses procedures and methods that produce acceptable crop yields and cause the least harm to the environment including man himself. The ultimate goals are optimum crop productivity and profit, without harmful alterations in the environment.

In contrast to pest *control,* pest *management* involves the planned continuous application of selected, integrated practices to maintain pest populations at or below levels that permit profitable crop production. Successful IPM requires the use of various combinations of resources (and weapons) that

Table 5.2.   A Useful Disease Management Program

- Select high-quality seed that is pesticide treated.
- Use resistant cultivars.
- Assay soil for nematodes and fumigate if necessary.
- Determine planting date based on weather forecast.
- Use proper row spacing and seeding rates.
- Apply pesticides properly at the correct rate and correct time.
- Harvest crop efficiently; store it properly.
- Destroy crop residues.
- Rotate crops.
- Keep abreast of latest developments.

assist and strengthen each other as components of the overall strategy in the war against pests. The farmer must assemble the necessary weapons and resources so that they are available in adequate amounts at the correct time and place (Table 5.2). Advance planning is necessary for production and deployment of such items as equipment, fertilizer, pesticides, seed, and water. Safe, efficient, and effective management practices must be carried out. The farmer needs to know what each weapon will do, how and under what circumstances it is to be used, and its relative effectiveness.

Economic Thresholds

The intelligent integration of disease management practices with crop management decisions is a vital part of today's agriculture. Before making a decision the farmer should ask: "Is a given management measure necessary? Will it pay? What risks am I incurring if I do not use a given management practice? How do the short-term gains compare to the long-term gains?" A necessary component of decision making is the establishment of *economic thresholds* of disease, i.e., the pest density at which management strategies and tactics should be used to minimize or prevent economic loss. Or, as the farmer asks: "Is this disease causing enough damage that I should spend time, money, and effort to manage it?" Pest losses must be based on measurements of both the quality and quantity of crop production. As a result of the rapid increase in computer technology, more extensive use of cost-benefit analysis of alternative pest management strategies and tactics is likely in the future.

Economic thresholds depend, in part, on the market value of crops. In deciding whether to plant a given crop in a given year, farmers must be aware of economic developments not only in their own community but in other countries. Market demands, government policies, and current production levels in other parts of the country and of the world all affect crop prices and, in turn, the level of disease infestation that warrants pest management efforts.

Sources of Information

"Be not the first by whom the new is tried, Nor yet the last to lay the old aside" might well be the motto for farmers in the complex world of modern agriculture. Successful farmers must undertake continuing education since every year sees a remarkable increase in the output of scientific data relevant to agriculture. The prompt acquisition and use of this information is one of the most vital components of farming, especially pest management, today. There is an unceasing supply of agricultural information available. Both public and private agencies make great efforts to assemble, digest, and dispense information to farmers and other interested citizens. Many countries now have regular programs on television and radio devoted to agriculture.

Newspapers publish agricultural sections, and numerous magazines are devoted to agriculture. Experiment stations, universities, and agricultural firms regularly publish articles, bulletins, circulars, handbooks, manuals, and periodicals. Television and radio programs on farm topics are broadcast daily in many communities. The extension services in some states have "teletips" that enable citizens to phone in and listen to a 3- to 5-min discussion on any of a large number of agricultural topics. Short courses, on-farm demonstrations, fairs, machinery exhibits, field days and tours are designed to convey the latest information to anyone who cares to come, see, and listen.

It is only common sense that alert farmers should seek out the latest information on efficient agricultural practices.

Agricultural Consultants

For more than 100 years, ever since the land grant colleges were established in the United States, county extension agents have provided invaluable service and given good advice to our farmers. County agents are trained in scientific agriculture and, as public employees, deliver and administer programs supported cooperatively by each state's agricultural extension service and the USDA. Traditionally, they have brought the latest information at no cost to farmers and advised them in the most efficient agricultural practices.

These extension agents seek out farmers to tell them of the newest discoveries. One reason agriculture is strong in the United States is because of the service of thousands of extension agents who have devoted their professional lives to helping farmers. However, there are too many farmers and not enough extension agents available to supply the continuing advice and information necessary to meet all individual farmer's needs.

Modern farm managers cannot hope to fight all their battles alone. They need individual help of personnel trained in the various tactics of pest management. Fortunately there is another source of help available—private agricultural consultants who hire or lease their expertise to farmers. These consultants perform such tasks as selecting proper cultivars, assaying the soil, monitoring pest populations, deciding where and if pesticides are needed, and correcting harvesting, storage, and marketing practices.

The agricultural consultant is a trained agronomist, economist, entomologist, pathologist, or member of a firm composed of skilled agriculturists who helps the grower plan a production program, giving advice on such diverse decisions as cultivar selection, planting dates, soil assays, monitoring for pathogen, insect and weed incidence, pest management practices, irrigation, harvest dates, proper storage, and marketing. Usually, these consulting firms charge on a per-hectare basis for their services and/or receive a bonus based on yield increases. Frequently, a farmer can pay for consultant services by omitting one unnecessary pesticide application based on information furnished by the consultant. As modern agriculture becomes more sophisticated, and farm enterprises increase in size and complexity, private consultants are assuming a vital role in helping agribusiness firms to avoid unnecessary risks and to be aware of the cost-benefit considerations involved with every decision made in growing a crop.

6

Management of Plant Diseases

Once the causal agent of a disease has been correctly identified, it becomes possible to develop plans to manage the disease. During the past 100 years, much research has been conducted (principally at land grant universities) on pathogens, the diseases they cause, and management methods. Today we can draw on this vast store of knowledge to help us in our war against plant pests.

Increasingly since World War II, new plant protection technologies have been developed and put into use. These new technologies have helped assure an adequate supply of food and fiber for the people of many countries and delayed the onset of famine. Now, however, the problem is more difficult. Not only must food production (along with adequate distribution) be increased, but the increase must occur despite growing economic, social, and political limitations on the technologies that may be employed.

Intelligent plant disease management is an economic necessity. Together with weed and insect management, it is a form of insurance that helps guarantee the farmer a profitable yield. It helps prevent wide fluctuations in yield. It prevents the destruction of lawns and ornamental plants. It helps prevent disastrous epidemics and famines.

There are at least five categories for the management of plant diseases:

1. Avoidance of pathogens by *exclusion* of the pathogen from a geographical area either voluntarily or by legislation and by *evasion* by the host plant to prevent the pathogen from coming into contact with the host.
2. Eradication or removal of the pathogen from the host, soil, from other reservoirs of inoculum, or from a geographical area. The actual result may be a reduction of inoculum rather than the total eradication of the pathogen.
3. Protection of plants by environmental changes or physical methods which produce conditions which are less favorable for disease development or by use of pesticides (see Chapter 7).
4. Use of cultivars with genetic resistance which resist the pathogen and disease development.
5. Chemotherapy with systemic chemicals to kill pathogens that are already in the plant (see Chapter 7).

Each of these approaches to disease management is discussed in the following sections or in Chapter 7.

Avoidance of the Pathogen
by Exclusion and Evasion

Many plants grow in areas where certain pathogens do not occur. When man upsets this pattern by moving plants, he may bring the plants into contact with a pathogen, and disease develops. Or man may move the pathogen on seed, transplants, planting material, or clothing to areas where the pathogen did not exist and where there are susceptible hosts. When a plant pathogen is transported to a new country or environment that permits optimum disease development on an intensively cultivated crop, great devastation often results.

One of the most effective ways to manage disease is to keep the pathogen away from the host (*exclusion*) or to keep the host away from the pathogen (*evasion*). This type of management can take various forms. A governmental unit (county, state or nation) can establish an embargo preventing the introduction of certain plants or plant products. These *quarantines* along with inspection are commonly used. In 1912, the U.S. Congress enacted the Federal Plant Quarantine Act. Among other things, this act prohibits the entry of plants into the United States. To enforce the act, inspection stations were established at all U.S. ports. Today it is virtually impossible to ship nursery stock from one country to another or from one state to another without the shipment being inspected and certified to be disease free. Plants may be grown in certain areas or in certain ways so that they can be guaranteed to be disease free.

Sometimes quarantines do not work. Airborne pathogens, for example can be blown hundreds of kilometers on the wind. Others are transmitted in or on seed. And sometimes people smuggle seed from one country to another. Some pathogens may be carried on clothing or in airplanes. In such cases it is difficult to stop the spread of plant pathogens.

If an area has a climate completely unsuited for survival of the pathogen or infection of the host, a quarantine may be unnecessary. Evasion—that is, growth of the host in areas free of the pathogen or under conditions unsuitable for the pathogen—may effectively contain plant diseases in some circumstances. For example, most bean seed in the United States is grown in the dry, irrigated regions of the West. Even though certain pathogens (especially bacteria) may be present, their spread and growth is restricted by the low humidities that prevail; as a result, the bean seed has less disease. Another example of evasion is time of planting. If cotton seeds are planted after the soil is warm, the seedlings can grow rapidly enough to evade the effects of several soilborne fungi that cause seedling diseases in cool wet soils.

Eradication of Pathogens

Sometimes it is impractical to try to exclude a pathogen from a given area. However there are other techniques that can be used. When a pathogen has been introduced into a new area despite a quarantine, or before quarantines have been established to prevent introduction, it is sometimes practical to eliminate or eradicate all infected host plants or those suspected of harboring plant pathogens. This helps prevent other plants from getting infected. Such eradication can occur across an entire state or region. Host eradication is also carried out routinely on individual plants in houses, greenhouses, home gardens, and farm fields to prevent the spread of numerous plant pathogens through elimination of the ready source of inoculum within the crop.

The practice of removing diseased, inferior, or abnormal plants from a field, greenhouse, or other planting is called *roguing*. This can be done by simply pulling out the plants, placing them in a bag, and then burning, burying, or otherwise disposing of them where they cannot spread inoculum. Roguing is widely used to help manage virus diseases. It is most efficient when performed early in the growing season when only a low percentage of plants are infected.

With other pathogens of annual crops—for example, cucumber mosaic virus (CMV), which overwinters mainly in perennial wild plants—eradication of the wild host sometimes eliminates the source of inoculum. Similarly, pathogens that require two hosts to complete their life cycle (e.g., cedar-apple rust) can be managed by eradication of the wild or less economically important host, which interrupts the life cycle.

Pathogens must survive from one growing season to the next if they are to cause damage. They survive in different ways or places called reservoirs. If these sources can be destroyed there will be no pathogen left to attack the host next year. There are many different kinds of inoculum reservoirs, each of which may require a different kind of sanitary measure to neutralize or destroy it.

Eradication is of little value against airborne pathogens because the wind will blow them back into the clean area. Eradication is most successful against pathogens that spread slowly and have a low rate of multiplication; soilborne pathogens are a good example. Only one disease has ever been totally eradicated. Careful roguing and inspections helped eliminate a bacterial disease of citrus from Florida. Attempts to eradicate other diseases have not succeeded.

Eradication from Infected Host Plants

Many animal diseases can be controlled by eliminating or eradicating the pathogen from a sick animal. The animal is cured. Not many plant diseases can be cured by eradicating the pathogen from a sick plant. The value of an individual plant determines if it is worth the expense to try to cure it. An individual bean plant may not be worth curing, whereas a large camellia bush or stately elm tree has great sentimental, aesthetic, and economic value.

Pathogens may be removed from sick trees or bushes by pruning off infected branches. Large wounds, those more than 8–10 cm in diameter, should be avoided to prevent them from becoming ports of entry for new pathogens. Earlier recommendations were to paint with shellac or to use wound dressings, but research has shown that painting does not protect the wounds from rot. Current recommendations are to make smooth cuts near the base of branches but not so close as to remove the tissue that aids in rapid healing of the wounds.

Bulbs, roots, seeds, and tubers are usually treated with chemicals to prevent decay after planting from seedborne pathogens or those occurring in the soil where they will be planted. Chemicals are applied as dusts or slurries, or by soaking the plant material in water solutions. Many seed merchants routinely treat seed before sale. The use of treated seed is an important part of the certified seed programs of several crops. Chemicals are available that penetrate into the developing seedling or remain in a zone around the seed and selectively kill or inhibit plant pathogens. Growers who plan to treat seeds should ask the local extension agent for the latest recommendations.

Plant materials, usually dormant buds, tubers, or woody tissues, may be heated to eradicate some pathogens. Certain viruses can be eliminated by heat treatment; full details should be obtained from the local extension agent. Propagative material can also be freed of certain fungi and viruses by use of meristem tissues or tip cultures. Carnations, chrysanthemums, and geraniums are usually propagated by cuttings. Oftentimes the top millimeter or so of the growing bud (*meristem*) is disease free. Slicing off this part of the bud with a sharp scalpel and growing it on selective media results in a disease-free plant, which can subsequently be used to obtain disease-free cuttings.

Some newer fungicides have systemic properties and can move through plant tissues to the pathogen and eliminate it from a plant. Some of these chemicals can now be used to cure some plant diseases (e.g., metalaxyl for late blight of potato).

Reduction of Inoculum

Sanitation

Sanitation practices provide conditions conducive to plant health and help to reduce the amount of inoculum available to start a disease outbreak. Another word for sanitation is cleanliness. Sanitation includes everything that can be done to reduce the amount of plant pathogens on a farm, in a field, greenhouse or storage house, in a group of plants, or even in a single house plant. By reducing the amount of inoculum, the chances of spreading the causal agent to other plants are lessened. Sanitation practices that remove infected fruit from packing sheds reduce the chances of spores of a pathogen falling on healthy fruit.

Burning

Burning of plant residues is a form of sanitation. For example, it is important to get rid of debris from the previous crop as soon as possible. An old practice that has been used for centuries is to burn the dried remains of the preceding crop. Thus, fire is set to the straw and stubble of grain crops. Spores, bacteria, and viruses that happen to be on or in the crop refuse are destroyed by the heat. Insects and weed seeds are also killed. Burning, however, has three disadvantages: it leads to wind and water erosion; the organic matter in the stubble is lost; and smoke from burning increases air pollution.

Controlled burning is also the primary means of managing the brown spot needle blight of longleaf pine (*Pinus palustris*). Vigorous longleaf seedlings in the grass stage (from the second year until they begin to elongate, about the sixth year) are very resistant to fire. Carefully controlled burns of low intensity will scorch the needles, and thereby kill the fungus without seriously damaging the seedlings.

Plowing or Disking

Another way to dispose of crop debris is to plow or disk it under as soon as the crop is harvested (Fig. 6.1). Spores and insects are buried, the crop refuse begins to rot quickly, and organic matter is added to the soil as the remains of the crop are recycled for use by the next crop.

Some soilborne pathogens are found principally in the top 15–30 cm of soil or subsist on the plant litter in the top layers of soil. If this upper, infested layer is inverted with a lower, noninfested layer, disease may be averted, especially if the host plants are shallow rooted. Special moldboard plows have been developed to invert the soil layers (Fig. 6.2).

Deep plowing is one of the principal methods recommended for control of the soilborne fungus *Sclerotium rolfsii,* which causes southern blight of peanuts and soybeans (see Chapter 12).

Deep plowing has some disadvantages. It is costly, it may disrupt the surface contour of fields, and in fields with shallow topsoils it brings infertile subsoil to the surface where it is difficult to manage.

Fig. 6.1. **Destroying remains of the previous crop reduces disease inoculum. The stalks are cut with a large rotary mower, then the roots are plowed out with a "middle-buster." After 2 to 3 weeks the land should be disked and sown to a small-grain cover crop to prevent water and wind erosion.**
Courtesy F. A. Todd

Without Hiller discs

B

With Hiller discs

C

Hiller discs

Fig. 6.2. Moldboard plows equipped with heavy-duty concave disk coulters are used to bury the surface layer of soil, containing shredded crop litter, 8 to 14 cm below the surface of the soil. *Courtesy C. W. Averre and S. F. Jenkins, Jr.*

Reduce-9-Pests. A multiple pest management concept using sanitation was introduced in North Carolina about 1960 by F. A. Todd. The practice, called R-9-P ("Reduce 9 Pests"), was a vital part of an integrated pest management program for tobacco growers. In the R-9-P approach, the simple job of stalk and root destruction by plowing and disking reduces root-knot, mosaic, vein-banding, and brown-spot infection; depresses hornworm, budworm, and fleabeetle populations; and helps control weeds and grasses. R-9-P is considered the last field job of the current crop and the first management practice for the next crop.

Plowing and disking of the roots exposes them to the drying action of the wind and sun, and the tops quickly die. Root-knot nematodes in the roots are killed, viruses are inactivated, spores are buried in the ground, insects are deprived of food sources, and weeds are destroyed. After 3 weeks farmers are urged to disk again and sow the fields with a cover crop (small grains) to prevent water and wind erosion. This practice causes rapid plant debris decomposition and recycles the plant waste for use by the next crop.

The value of the R-9-P program depends upon the timing, the quality of the job completed, and the number of growers participating. Because early destruction of stalks and roots is vital, TV and radio announcements are broadcast urging farmers to begin disking the same day that they finish harvesting the crop. A high-quality R-9-P program increases the effectiveness of other pest management tactics, including the use of resistant cultivars, rotations, and chemicals.

R-9-P is a statewide program in North Carolina. Therefore, it reduces inoculum across the entire state. To get 100% participation in all counties, farmers elect R-9-P Committees in each township to oversee the program and help insure participation by all farmers. A sense of community pride is developed, and everyone benefits from this sound husbandry. Commercial chemical companies donate prizes to each community with 100% participation.

Soil and Root Media Pasteurization

One of the first rules of agriculture is to get the crop off to a good start by using healthy seed and/or disease-free transplants. Therefore, if at all feasible, growers should sow crops in pathogen-free soil. For many low-value, directly seeded field crops it is seldom feasible to eradicate pathogens from the soil. However, eradication of pathogens is practical and widely done in seedbeds used for growing transplants of high-value crops (e.g., flowers and bulbs), as well as in golf greens and greenhouse soils. Heat, in the form of steam, is an efficient method of pasteurizing soil. Steam penetrates soil quickly and imparts a large quantity of heat without raising the soil temperature excessively.

The term *pasteurization* is preferred over *sterilization*. Sterilization refers to the destruction of all organisms in root media, whereas pasteurization indicates that only selected organisms are destroyed. Successful pasteurization eliminates the harmful soil pests but does not kill all beneficial organisms, which rapidly build up to normal levels after pasteurization. If a disease organism is the first to reinfest a pasteurized medium, however, it also will develop rapidly without competition and may cause much damage to plants in the medium.

Greenhouse Soils.   Pasteurization of soil and other root media to control pathogens, insects, and disease is a standard practice for greenhouse crops. It is generally done annually or between crops. This is necessary because of the threat of a rapid buildup of disease in greenhouses. In cold regions, summer is the preferred time for pasteurization because crop production is low, summer labor is available, and outside temperatures are higher, so that less steam is required and sufficient boiler capacity is available.

Pasteurization procedures.   Each of the soil-inhabiting pests that injure plants has a temperature lethal to it (Fig. 6.3). Usually soil pasteurization is completed when the temperature in the coldest part of the soil has remained

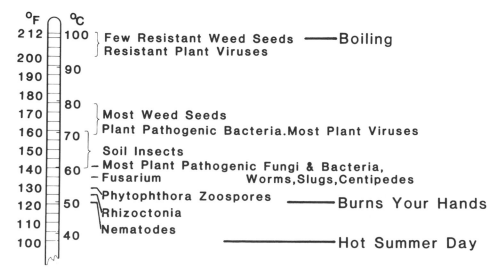

Fig. 6.3. Temperature necessary to kill pathogens and other plant pests. Most of the temperatures are for 30 min exposure under moist conditions.
Adapted from K. F. Baker

for 30 min at 82°C. At this temperature most harmful organisms are destroyed, whereas many beneficial organisms are not killed.

The growth medium should be loosened before pasteurization. The soil should not be dry but in good working condition. If it is too wet, more heat will be required to attain the proper temperature. Water requires five times as much heat to raise its temperature as does soil with proper aeration.

If there are many weed seeds in the soil, wet the soil a week or two before pasteurization. The weed seeds will begin to germinate and will be killed at lower temperatures. Add fertilizers, limestone, peat moss, bark or other amendments to the media as needed and mix well before pasteurization.

Usually the steam boiler used for heating a greenhouse can be used for pasteurization. Portable steam generators are also available. Steam is distributed in ground beds through buried perforated pipes or tile. Raised benches filled with root medium may be pasteurized with or without buried pipes, but a cover is necessary. If pipes are used they are buried at half the soil depth. Another method is to inject steam between the cover and the root medium through canvas hoses.

Potting media can be pasteurized in a wagon equipped with perforated steam pipes at the bottom (Fig. 6.4). A few pots of soil can be pasteurized by heating them in a stove or oven. Pots and flats can be pasteurized in cabinets (Fig. 6.5) equipped with a steam outlet.

Covers are placed over the soil media during pasteurization to catch the steam and hold in the heat. Polyethylene, vinyl, or nylon covers are commonly used. Polyethylene is cheapest but lasts only 1 year. Nylon is the most expensive but can be reused at least 100 times.

Fig. 6.4. **Pasteurizing soil. Steam from a boiler enters the aerator (foreground), while a blower introduces air. The two gases mix and are then conducted through a hose to a covered potting wagon (background) containing the soil. A hand valve regulates the amount of air introduced, which in turn controls the temperature. With this device, soil can be pasteurized at 60°C rather than 100°C.**

Fig. 6.5. **Sterilizing pots and flats in a steam cabinet using steam from a greenhouse boiler. The cabinet doors are closed during the actual sterilizing process.**

After-steaming problems. Large quantities of manganese exist in many soils. High levels of manganese are toxic to some plants and also interfere with uptake of iron by roots. Iron deficiency is commonly associated with high manganese levels. Fortunately, only a small but adequate amount of manganese is available for plant use in most soils; the majority of it is in a form unusable by plants. Steam pasteurization, however, converts more manganese to a plant-available form. Thus, the longer soil is steamed, the greater the risk of manganese toxicity.

Root media that contain organic matter rich in nitrogen can release ammonium nitrogen during and after pasteurization. (Such media include compost, leaf mold, manure, and peat.) Normally, ammonium nitrogen is continuously converted to nitrate nitrogen by soil bacteria. During pasteurization, ammonifying and nitrifying bacteria are nearly eliminated and it may take 3–6 weeks before populations build up to normal. During this time toxic quantities of ammonia may develop. Root tips may be injured and the plants remain stunted and wilted. Ammonium toxicity can be avoided by not overpasteurizing and by delaying use of the media until 2 weeks or more after pasteurization.

Field Soils. Limited areas of field soil can be steam-pasteurized and this is commonly done, particularly where methyl bromide, a chemical commonly used to fumigate soil, is not suitable or available. The soil should be well prepared and not too wet. The boiler may be fixed in a central location or may be mounted on a truck.

Steam is conducted from the boiler by a hose to a steam rake, which consists of a pipe header (10 cm in diameter and 4 m long) that is drawn by a cable across the field. Projecting into the soil from the header are chisels, 45 cm long and spaced 25 cm apart. Behind each chisel is a 1-cm pipe that releases the steam into the soil. A winch is often used to pull the rake across the field at a rate about 50–60 cm/min. A pasteurizing cover, sufficiently long to require 30 min to pass over any given point, is attached behind the header. The cover holds the steam in the soil and helps maintain the temperature at or above 70°C for at least 30 min. One hectare of soil can be pasteurized by a single rake in 80–140 hr.

Storage House Disinfestation

Many crops must be stored for varying lengths of time before they can be sold. Often the storage house becomes contaminated with rot-producing organisms that survive between successive crops on the walls and floor of the storage house and on crates. If these organisms are not killed, they may start rotting the new crop. An important disease management measure is to thoroughly clean the storage house and surroundings of all crop remains, dirt, and other trash. The house should then be washed down with some disinfectant such as household bleach (sodium hypochlorite), according to directions on the label. Advice may be obtained from the local county agent.

Refuse Site Destruction

Many crops must be sorted, graded, and/or trimmed before they are sold. Imperfect, diseased, bruised, or improperly sized produce, in particular, must be discarded. Sometimes large piles of rejected plant material accumulate. These cull piles serve as an inoculum reservoir for plant pathogens. For example, potato farmers often discard or throw away tubers that are discolored or rotten. The potatoes in such cull piles may be infected with the late-blight fungus. In the spring some of these infected tubers sprout and the fungus sporulates or fruits on the new growth. The spores are blown to potato plants growing in nearby fields and start new infections or outbreaks of late blight. Destroying these cull piles by spreading them on the land and plowing them under, by burning, or by spraying with a herbicide can eliminate the primary inoculum for this disease. The damage from late blight is in part related to the time of infection—the earlier the infection, the more damage done. Thus, anything that delays the start of infection will help decrease losses.

Disinfestation of Machinery and Tools

Several plant pathogens can survive on or are spread by machinery and tools. For example, when Irish potato tubers infected with the bacterium that causes ring rot are cut to make seedpieces, the cutting knives can become contaminated and spread bacteria to healthy tubers. A few infected tubers may lead to contamination of all the seedpieces. Therefore, knives used for cutting seedpieces should be periodically disinfested with sodium hypochlorite, and contaminated containers disinfested by dipping in formaldehyde or sodium hypochlorite before use.

Tobacco mosaic virus (TMV) is in the juice of infected tobacco, tomato, and pepper plants. If such plants are bruised or cut by equipment used for cultivating, pruning, suckering, or harvesting, the virus can be spread to and infect healthy plants. This commonly occurs in greenhouse crops and causes great losses. This spread can be prevented by having workers dip their hands or pruning shears in milk after handling each plant.

Cropping Systems or Rotations

The practice of growing the same crop in the same field year after year is called *monoculture*. Perennial crops such as apples, peaches, grapes, forest trees, legumes, certain pasture grasses, and turf grasses occupy the same fields for many years. However, some farmers choose to grow annual crops such as corn, wheat, rice, soybeans, or tobacco in the same fields year after year. Farmers who practice monoculture must be constantly on guard against disease outbreaks, for continuous presence of a host may permit pathogens to build up to high levels. Epidemics then occur, and an orchard, vineyard, or forest stand may have to be replaced or an annual crop may be lost.

The practice of growing different crops successively in the same field is called a *cropping system, cropping sequence* or *rotation*. The cropping sequence may be completed in 1 year or extend over several years. For example, in frost-free areas or regions with long growing seasons, it is possible to plant annual crops the year-round or grow two or more crops in the same calendar year. Thus in California, Florida, and Texas, certain vegetable farmers alternately grow crops of tomatoes and sweet corn. In parts of the southeastern United States, it is a common practice to plant soybeans after small grains. In cold regions where crops cannot be grown the year-round, a cropping sequence may take 4 years to complete. In the northeastern United States, a common 4-year rotation on dairy farms is to grow a crop of corn followed by oats the next year, then wheat, then hay in succeeding years. In the midwestern United States, alternate crops of corn and soybeans compose a 2-year rotation.

Crop rotation (Fig. 6.6) is an efficient and economical weapon for controlling plant diseases. It is a very old method of farming: Vergil, the Roman poet who lived in 30 B.C., praised the use of crop rotations.

Fig. 6.6. Control of bacterial wilt by crop rotation. A—Planted in tobacco continuously. B—Tobacco following 3 years of corn.
Courtesy USDA

Many farmers, however, cannot or will not use long cropping sequences of 4 years or more. They don't have enough land or they prefer to risk the chance of disease rather than use their best fields only 1 year out of 4 for their principal cash crop. Some rotations are impracticable, as when certain fields cannot be irrigated, or are inaccessible, or are too steep or stony or unsuitable for a given crop in a sequence. Certain rotations may be uneconomical: allowing land to grow up in weeds for 2 years or more will help control root knot, but it is less profitable than growing cultivated crops. Bare fallowing, the practice of periodically plowing or disking land during the growing season without seeding it, for the purpose of destroying weeds and conserving moisture, also reduces the incidence of some diseases and can be used in a cropping system. But it is expensive to keep the land disked. Furthermore, the land is subject to erosion and the farmer loses use of the field during the fallowing period.

As diseases continue to spread, many fields become infested with more than one pathogen. Therefore the planning of suitable and effective rotations is more difficult on some farms than others. In those fields where more than one disease is present, the farmer must determine which disease is causing the most trouble and plan the rotation to combat this disease, even though such a rotation may produce conditions favorable for "minor" diseases.

Many factors must be considered in planning a satisfactory cropping system for a given farm or given region. Although it is difficult to outline any one cropping sequence to be used under a particular set of conditions, it is possible to list the factors that influence the choice of a cropping sequence. A good cropping system should have the following characteristics:

- Be profitable
- Maintain the proper soil productivity, keep a desirable microbiological balance, and utilize fertility efficiently
- Help manage weeds, insects, and diseases
- Maintain the soil in good physical condition
- Reduce wind and water erosion
- Be flexible, practical, and easy to follow
- Include both row and sod crops

The effectiveness of disease management by crop rotation depends upon many factors but principally upon the life history and behavior of the pathogen, including (1) the persistence of the pathogen in the soil, (2) the rate of buildup of inoculum, (3) the means of spread of the pathogen, and (4) the degree of host specificity. In relation to the soil, plant pathogens may be classified as soil invaders or soil inhabitors.

Soil invaders include those pathogens that usually attack the aerial parts of plants but may reside temporarily in crop refuse or in surface layers of soil. The use of short rotations may be of great value in reducing the damage from these diseases. Included in this group are a number of leaf-spotting fungi such as *Alternaria, Cercospora,* and *Colletotrichum.*

Soil inhabitors are those pathogens that reside naturally in the soil. *Persistent* soil inhabitors are capable of indefinite existence in the absence of the host. Some soil inhabitors persist by means of long-lived, resistant spores or resting bodies. These organisms may be quite host specific. Long rotations may reduce the inoculum level of these organisms but will generally not eliminate them. Examples of this type of soil inhabitor are fungi belonging to the genera *Fusarium* and *Phytophthora*. A second group of soil inhabitors persist because they are well adjusted to the soil environment and can live as saprophytes in competition with other soil organisms. Many of them can persist indefinitely in the absence of the host plant, and they may have a wide host range. Examples are pathogens that cause root and stem rots, seed decay, and seedling damping-off, such as *Pythium* and *Rhizoctonia*. Crop rotation is of little value in managing them.

Nonpersistent soil inhabitors are those pathogens found in the soil that can develop and reproduce only in the roots of suitable host plants. They are sometimes called *root inhabitors* and include various species of plant-parasitic nematodes. Rotations involving the use of nonhost plants are useful in management of these pathogens.

Some general principles about crop rotation can be summarized as follows:

- Setting up a successful disease-management rotation involves (1) identification of the disease, (2) selection of crops that are adapted to the farm and locality, and (3) effective arrangement of crop order in the rotation.
- In general, longer rotations are more effective in managing diseases than short rotations, and short rotations are better than no rotation.
- Specific crop rotations are not equally effective with all diseases.
- The sequence of crops in a rotation should not be rigid and inflexible; the crops in the rotation should be changed as often as practicable.
- A good rotation crop may lose its value, if grown too often, by permitting the buildup of populations or strains of pathogenic organisms that will attack it and other crops in the rotation.
- Rotations that affect the carry-over of soilborne pathogens have other important effects on crop yields; thus, figures for crop yield and amount of disease are not always closely parallel. Those rotations that are favorable from the nutritional viewpoint may sometimes give high yields despite extensive development of disease.
- The proper use of cover crops is a valuable part of crop rotation. They prevent the loss of plant nutrients by leaching, washing, or blowing of the soil. In preventing the movement of soil by wind and water, cover crops reduce the spread of soilborne pathogens.

Cropping sequence alone is not the answer to disease management. However, when it is intelligently combined with soil fumigation, the use of resistant cultivars, and proper cultural practices (soil conservation, correct plowing, sanitation), disease losses will be minimized. With diseases as prevalent and destructive as they are in many areas, no grower should plant his principal crop year after year in the same field unless there is no alternative. In planning an overall pest management system designed to fit his own particular operation, each grower should consider carefully the benefits and/or limitations of rotation in his production program. In addition to the benefits to be derived from disease management, rotation of crops helps to prevent wind and water erosion, supplies organic matter to the soil, and helps maintain fertility and proper soil structure.

The applications of crop rotation in the management of specific diseases are discussed in the chapters dealing with each disease.

Greenhouse Management

Disease prevention and sanitation are especially important in greenhouses, which are designed to provide optimum conditions for plant growth. This subtropical environment of warm temperature, high RH, and adequate water is also conducive to the growth of many disease organisms. In addition, continuous culture of the same or related crops the year-round exacerbates disease problems by providing an ever present host on which pathogens can multiply.

When high populations of plants (intensive cultivation) grow in a limited area of land or space, the chances of disease are increased. Once a disease gets started in a greenhouse, it is difficult to manage because conditions are so favorable for development of epidemics. In such cases it may become necessary to discard and destroy an entire house full of plants; in addition to the direct loss, the cleanup job is laborious and expensive.

Prevention plays an important role in disease management in greenhouses. Disease organisms may be carried in the soil and water, on pots and tools, on shoe soles and clothing. Windblown spores easily invade a greenhouse. Since they are invisible, their presence is known only when lesions or other symptoms become visible.

Diseases are often the downfall of the ignorant or careless greenhouse manager. Careful growers realize that disease is inevitable. Therefore, they are continually on guard against them. An efficient disease management program is just as necessary as daily watering in greenhouse production.

A number of basic do's and don'ts can go a long way toward minimizing plant diseases in greenhouses:

- Use root media pasteurized with steam or methyl bromide.
- Use sterilized pots, containers, and equipment. All pots, flats, and plant supports should be sterilized before reuse.
- Sterilize the potting bench and keep it clean. Sweep after use with a special broom kept only for this purpose. Swab off once a week with disinfectant. Do not put contaminated potting media or pots on the potting bench.
- Store potting media in special bins. After potting media have been pasteurized, they are easily recontaminated. Do not sweep soil from the floor into storage bins. Prevent dust from blowing from one bin to another.
- Avoid contaminated soil. Take all possible precautions against carrying foreign soil into greenhouses. Keep a flat pan containing a disinfectant solution at the door so that visitors can step in this to sterilize their shoes; or have visitors use disposable overshoes. Do not place watering hoses on the floor where they can become contaminated; instead, fasten a clip to the side of the bench to hold hoses between watering.
- Do not water carelessly or hurriedly. Avoid splashing water from one pot

to another. Do not overwater, as root-rotting organisms thrive in wet soils. Use automatic watering devices whenever possible. Construct walkways so that water puddles cannot form. They can harbor pathogens.

• Clean up plant debris. Do not let weeds grow under the benches. Do not throw poorly rooted cuttings under a bench. When disbudding or pruning, use a basket or bucket to hold plant parts till they can be carried to an incinerator. Workers handling tomato or pepper plants should dip their hands in milk before moving from one plant to the next to prevent spreading virus diseases. Impress upon employees that cleanliness is imperative. Devise sanitation practices that employees can follow. Think clean! Establish trash dumps or compost heaps far enough from greenhouses so that spores, insects, or seeds cannot easily be blown back into greenhouses.

• Do not keep leftover plants after the season in the hope that they may be sold. Dispose of the preceding crop as soon as possible so that it will not be a source of infection for the next crop.

• Use clean stock. Purchase bulbs, cuttings, or seedlings from reputable sources. Discard any weak or diseased materials. Use only certified seed. Remember that TMV can be carried on tomato seed. Take cuttings for propagation only from healthy plants. Periodically sterilize the cutting knife by dipping in alcohol and flaming. Work on a clean tabletop or clean newspapers.

• Maintain proper conditions. Many plant pathogens thrive when humidity is high, and most fungus spores require free water to germinate. Thus, steps should be taken to control humidity and condensation. Warm air holds more water than cold air; as air cools, its moisture-holding capacity drops until the dew point is reached and water (condensation) forms on any solid surface. Condensation can be combatted by opening ventilators for 15–20 min when the heat first comes on to allow warm water-filled air to escape. After the drier, incoming cold air is heated, the ventilators are closed. The relative humidity can also be reduced by watering early in the day; as the air warms it will then gradually pick up moisture.

• Develop a regular spray program for management of diseases, insects, and mites, and follow it carefully. Damping-off and root-rot diseases can be inhibited by drenching soil with fungicides. Leaf- and flower-spotting organisms—including mildews, rusts, and leaf spots—can also be prevented by regular application of fungicides. Precise recommendations should be obtained from a state experiment station.

• Greenhouse managers should subscribe to and read one or more trade publications to keep aware of new developments. Conferences and training seminars also are helpful, as are gardening programs on radio and TV, especially for nonprofessionals with home greenhouses. Visits to other greenhouses also can uncover effective practices.

Weed Management

Efficient weed management is a vital component of an efficient pest management program. Weeds harbor plant pathogens and insects and provide hiding places for rats and other rodents. Uncontrolled weeds indicate a careless grower. More importantly, weeds reduce profits by using water and fertilizer that could be used by the crop, by interfering with harvesting machinery, and by encouraging pests.

Fence rows, lanes, and the farmstead should be kept mowed and free of weeds. Areas around greenhouses, cold frames, and seedbeds should be kept weed free. When the transplanting season is over, seedbeds should be disked promptly, to destroy all unused seedlings, and sowed to a cover crop such as soybeans or other legumes. Growers should become familiar with weed management practices recommended in their state.

Insect Management

Insect management is a necessary part of an integrated pest management program. Almost everyone is familiar with insect damage to plants such as that caused by the army worm, boll weevil, the ear worm, Japanese beetle, scale insects, wire worm, and many more. Aside from the direct damage resulting from insect feeding or egg laying, insects carry plant pathogens (see Table 4.1). The tiny holes made by insects provide entry points (*infection courts*) for microorganisms; oftentimes insects carry the pathogens that enter through the wounds.

The same principles that apply to disease management also apply to insect management (see Table 5.1). Therefore, it is necessary for growers to have an overall pest management strategy in order to keep losses to a minimum.

Protection of Plants by Adjustment of the Environment

Farmers have learned to manipulate the environment by such practices as varying the time of planting to avoid disease, shading the soil by mulching or spacing of plants, and using windbreaks to reduce damage from winds. Any farming practice aimed at improving the vigor of the plant often helps to increase its resistance to disease. Therefore, as part of their overall efforts to manage plant diseases, farmers should make every sensible effort to grow a healthy crop.

Fields should be prepared properly; crops should be planted at the proper time and properly fertilized; seeds should be planted at the proper depth, with proper spacing between plants. Weeds and insects should be managed. Good soil drainage and proper irrigation should be used. Sprinkler irrigation is widely used to prevent frost damage. Erosion should be prevented by use of terraces, contoured rows, and strip cropping with sod and row crops. Soil pH should be maintained at the proper levels. Proper harvesting and storage are essential; it is a waste to grow a bumper harvest and then lose it with careless harvesting or storage.

Protection of Plants
by Physical Methods

Sometimes it is impracticable or impossible to exclude or eradicate a pathogen from a given area. In such cases the crop must be protected from the pathogen by either physical or chemical means. Physical means, which are discussed in this section, include cooling, moisture management, drying, and radiation. The management of plant diseases by chemicals is discussed in Chapter 7.

Refrigeration

The most widely used method of preventing postharvest diseases of fleshy or succulent plant products is refrigeration. Temperatures of 0°–5°C inhibit the growth of many pathogenic microorganisms, thus preventing the spread of existing infections and the start of new ones. Most perishable farm products are cooled immediately after harvest, transported in refrigerated trucks, railroad cars, or airplanes and kept refrigerated until used by the consumer. One reason the U.S. citizen can eat fresh fruit and vegetables year-round is the wide use of refrigeration by commercial companies and in the home.

Moisture

High moisture levels in air and soil are conducive to many plant diseases. Some crops are regularly grown in low-humidity dry areas and surface-irrigated to reduce or prevent certain leaf-spot diseases. By keeping the leaves dry, farmers can prevent infection. Certain root-rotting fungi thrive in wet soils. The damage they cause can be reduced by not growing crops on waterlogged soils. Or plants can be grown on elevated beds and surface-irrigated between the beds. In special cases (golf greens and greenhouses) soil mixtures with good drainage are used.

Drying

The moisture content of crops such as corn, peanuts, sorghum, soybeans, and wheat must be carefully regulated if damage by storage fungi is to be avoided. These crops should be harvested at the proper stage of maturity and at the proper moisture content. Then they should be dried properly to the correct moisture content for storage. In many of the grain-producing areas of the world, grain is harvested in the dry season or during sunny periods. However, if wet weather prevails and harvest is delayed or the grain cannot be dried properly after harvest, storage rots can markedly reduce the quality of the crop. Barley, corn, rice, and wheat should be stored below 13% moisture and soybeans below 12% moisture. As temperatures decrease, fungus growth decreases. Storage temperatures below 10°C decrease the rate of deterioration.

Fungi are a major cause of damage, spoilage, and heating of stored grains. Storage fungi cause loss of germination, bin-burning, bad odors, mustiness, and heating. Moldy grain makes feed less palatable and nutritious for live-stock, and renders some foods unsuitable for human consumption. Some of the fungi that grow in grains and other seeds before harvest or during storage produce *mycotoxins* (mold-produced toxins). The common storage fungus, *Aspergillus flavus,* produces several toxins that are called *aflatoxins.* These toxic chemicals (*metabolites*) are produced by several fungi when they grow on corn, cottonseed, peanuts, and many other substances. Feeds contaminated with aflatoxins are toxic to cattle, ducklings, swine, and turkeys. Younger animals are more susceptible to injury than older animals. Only a few parts of aflatoxin per billion can cause pathological changes in susceptible animals. Therefore, it is risky to feed moldy grain to farm animals. It may make them sick, reduce weight gains, or even kill them.

Aflatoxins can be reduced or prevented from occurring in corn if farmers grow adapted cultivars, harvest the crop when mature, dry it promptly, and store the grain properly. Similar procedures should be used when harvesting other crops. Precise instructions can be obtained from local extension service personnel.

Sun drying of certain crops is still practiced in many parts of the world by small farmers. Grapes are dried to make raisins, plums for prunes; coffee berries, dates, and figs are dried in the sun. Certain types of tobacco are sun dried; apples or peaches are sliced and spread in the sun to dry. Needless to say, once sun-dried fruit absorbs excess moisture it quickly spoils at warm temperatures.

Radiation

Ultraviolet (UV) light, X-rays, gamma rays, and particulate radiation such as alpha and beta particles destroy microorganisms and can be used to free plant material of pathogens. Unfortunately, the dosage necessary to kill pathogens also injures plant tissues. But we may expect a breakthrough anytime; some day refrigerators may be as obsolete as food-storage cabinets in the home.

Disease-Resistant Cultivars

For many crops disease-resistant cultivars are the least expensive, safest, and most practical way to combat most diseases. The use of resistant cultivars is particularly appealing to those who must rely on expensive, frequent pesticide applications to protect large acreages of low-income crops. As crop production costs increase, growers must diligently seek ways of reducing these costs. The elimination or reduction in pesticide use would help do this. Also the use of resistant cultivars requires few or no changes in farmers' production practices; all they need do is substitute seed of a resistant cultivar for that of a nonresistant one. Often the cost of the seed is the same. However, it does take skilled scientists, time, and money to breed resistant cultivars.

The early phases in the development of a disease-resistant cultivar are usually tedious and expensive. In the United States, much of this work is performed by state experiment stations and land grant universities that are supported by tax dollars. Recent changes in patent laws have made it feasible for many private seed companies to employ plant breeders to develop resistant cultivars. These private breeders use breeding lines or *germplasm* sources developed in state experiment stations, USDA research centers, and their own facilities to build new cultivars for release to growers. The production of reliable resistant cultivars requires the cooperation of both public and private agencies. It is safe to say that without this cooperation the use of disease-resistant cultivars would have lagged far behind what it is today and crop yields would be much lower than they are.

Disease resistance in crops is the foundation of a disease management program. But resistant cultivars do not constitute the complete answer to disease problems because they are not available for all crops. Moreover, the use of disease-resistant cultivars has its disadvantages. Cultivars adapted to one locality may not do well in another. Seemingly unimportant differences in environment may make the difference between a high-grade and low-grade crop. Specific cultivars may respond differently to rainfall, the amount and kind of fertilizer, soil type, planting density, day length, length of growing season, and cultivation practices.

The use of disease-resistant cultivars also may lead to problems stemming from the genetic uniformity of such cultivars and the tendency for new strains of pathogens to develop or for a minor pathogen to become an important one.

Genetic Uniformity and Plant Diseases

When a new cultivar or several closely related ones resistant to only one strain (race) of a pathogen are introduced and planted over a wide area, rapid development and spread of other races of the pathogen that will attack the new cultivar frequently occur. As a general rule for annual crops, a resistant cultivar lasts about 5 years. Then it "breaks down" or wears out. Sooner or later an epidemic will develop and great losses result. This is what happened when southern leaf blight struck the corn crop in 1970. Genetic uniformity in crops, although desirable agronomically, is epidemiologically unsound and often catastrophic. Therefore, disease-resistant cultivars should be used in conjunction with other disease management measures.

Another example of the dangers of genetically uniform resistant cultivars— putting all your eggs in one basket, so to speak—is the Victoria-based cultivars of oats. Victoria was developed with resistance to crown rust and smut, two important oat diseases in the Midwest; it also gave good yields, had short stiff straw, and did not lodge easily, making it easy to harvest with a combine. Breeders in the United States used Victoria as a parent in developing some 30 oat cultivars. In the mid-1940s approximately 75 million ha of oats—two-thirds of the U.S. production—were planted to these cultivars. However, in 1944 a strain of the *Helminthosporium* fungus caused much damage to Victoria and related cultivars. In 1946 the epidemic of Victoria blight caused such severe losses to oat cultivars with Victoria breeding in their makeup that they had to be abandoned.

Sometimes, a cultivar resistant to a major disease turns out to be susceptible to a pathogen that usually causes only minor damage. After the resistant cultivar is planted over a wide area, the secondary pathogen may increase and become a major problem. This type of situation occurred after the blue mold (downy mildew) epidemic of tobacco caused a $25 million loss to the European tobacco crop in 1960. European breeders, in desperation, urgently requested and obtained seed of blue mold-resistant tobacco cultivars from other countries. Before these blue mold-resistant cultivars were introduced into Europe in the early 1960s, potato virus Y (PVY), the causal agent of vein banding, was a minor pathogen. Most of the European tobacco cultivars were tolerant to this virus and vein banding caused only minor losses. However, the blue mold-resistant cultivars that were introduced and widely used since then are susceptible to certain strains of PVY. Vein banding became a serious tobacco disease in Europe, and tobacco breeders are urgently seeking ways to control this threat to the tobacco crop.

It is difficult enough to breed a usable cultivar resistant to one disease; it is increasingly difficult to incorporate resistance to two or more diseases in one cultivar. Occasionally, this feat has been accomplished. One of the most widely grown tobacco cultivars in the United States is resistant to four diseases and tolerant to a fifth.

Kinds of Disease Resistance

Some plants are *immune* to a particular pathogen. That is, the pathogen cannot cause disease even under favorable conditions. Other plants exhibit certain degrees of resistance ranging from near immunity to complete susceptibility. Tolerant cultivars can produce a good yield despite infection, whereas susceptible cultivars are greatly damaged or killed by the pathogen.

The ability of a plant to resist disease is an inherited characteristic. It is controlled genetically. Plants undergo sexual recombination and change genetically. The ability of a pathogen to cause disease is also inherited and controlled genetically. Pathogens also change genetically. Some pathogens are composed of many strains, or races, each differing in its ability to attack a given host cultivar. Disease expression in each host-pathogen combination is predetermined by the genetic material of the host and the pathogen. Cultivars resistant to only one pathogenic strain possess *race specific* or *vertical resistance* (VR) genes, whereas cultivars resistant to many pathogenic strains possess *nonspecific, horizontal,* or *general resistance* (HR) genes.

General resistance usually is only moderate (i.e., cultivars with HR may be damaged to some extent by a pathogen but not as much as similar cultivars lacking the HR genes). Cultivars with VR generally exhibit a higher degree of resistance than do those with HR, and they are often easier to develop in a breeding program. However, it is more likely for new races of a pathogen to arise in response to a VR cultivar than to an HR cultivar. Thus, cultivars with VR tend to break down faster than those with HR (see discussion in previous section).

Any heritable characteristic that helps localize and isolate the pathogen at the point of entry (*infection court*), that reduces or neutralizes the harmful effects of toxic substances produced by the pathogen, or that inhibits pathogen growth contributes toward host resistance. Any characteristic that helps the plant grow and mature under conditions unfavorable for the pathogen also contributes to resistance. On the other hand, any characteristic that induces changes in the host that increase growth of the pathogen or the production of toxins harmful to the host, or that benefits the pathogen in any other way, contributes to susceptibility. It must be remembered that disease is the result of the interaction of a multitude of chemical processes initiated by both the host and pathogen while in contact in a favorable environment. The genes of the host plant that control disease resistance do so by changing or adapting the physiological processes of the plant so that infection or disease development is neutralized or prevented from operating.

When the pathogen and host come in contact, a battle begins. If the pathogen wins, disease develops. If the host wins, there is little or no disease.

Developing Disease-Resistant Cultivars

There are four steps in breeding a disease resistant cultivar: (1) obtaining a source of resistance, (2) transferring the resistance to a breeding line, (3) evaluating the lines for disease resistance, yield, and quality, and (4) releasing seeds or plant material to growers.

Sources of Resistance

All plants growing on the earth today are the result of natural selection and breeding of plants that evolved in different kinds of climates over millions of years. This evolution has produced countless genetically different forms, both wild and cultivated. The fact that these plants have survived indicates that some of them possess numerous genes for disease resistance.

An easy and quick way to obtain resistance is to select resistant plants from a cultivar already in use, save seed from it, and use this seed for the next crop. Sometimes this process is not possible. Then it is necessary to search wild or related species of the crop in question to find a source of resistance. This requires a diligent search of seed stocks and exhaustive testing to determine the quantity, method of inheritance, and type of resistance available. This approach requires that stocks of viable seed of all cultivars and wild species be maintained and available for future study. Seed collections of most of our crop plants are maintained by the USDA, by state experiment stations, and by universities and some seed companies. Scientists in other countries also maintain valuable seed stocks.

Transferring Resistance

Once a source of resistance is found, the next step—a big one—is to transfer this resistance to a breeding line by hybridization. Crosses are made between resistant plants and susceptible plants and the offspring exposed to the pathogen so that resistant plants can be identified and saved. Sometimes successive crosses (*backcrossing*) must be made. The success of the whole program depends upon maintaining and keeping only those individual plants that are disease resistant and have desirable agronomic characteristics such as high yield and good quality.

One big disadvantage of the backcross method is that it takes 5–10 yr or longer to incorporate or stabilize resistance, particularly to several diseases, in cultivars that also have acceptable yield and quality. Much time can be saved by the use of F_1 hybrids. In this method, a breeding line with dominant resistance is crossed with a closely related, agronomically desirable line; the seed from this cross is planted to produce a crop made up of a uniform population. Most of our corn cultivars are F_1 hybrids. The disadvantage of the F_1 method is that hybrid seed does not breed true; hybrid seed can only be obtained by special crossing procedures.

There are other genetic techniques that can be used to incorporate resistance into breeding lines, and new ones are being discovered. Advances in biotechnology will help develop resistant cultivars more quickly in the future (see Chapter 8).

Evaluation

The third step in breeding disease-resistant cultivars is evaluating the lines in which disease resistance has been incorporated. This is time-consuming, laborious, and often expensive. Evaluation can take place in the greenhouse or in the field. During the backcrossing phase of the breeding program, when resistance is being stabilized into an acceptable cultivar, a greenhouse is almost indispensable. Since many plants will grow and flower year-round in a greenhouse, two or three generations (crops) may be grown annually. With suitable inoculation techniques and adequately controlled greenhouse conditions, successive backcross generations can be exposed to pathogens and the susceptible individuals eliminated in a reasonably short time.

Next comes field testing, which must be done to avoid release of unacceptable or defective cultivars. After resistance has been stabilized, the various lines are grown in field plots. These field plots usually are of two types: disease free or disease infested. In the disease-free plots, the lines are tested for yield and quality in comparison with standard cultivars. In the disease-infested plots, the cultivars are tested for disease resistance under field conditions. Field plots may be located several hundred kilometers apart, in frost-free areas or tropical regions, to speed up the testing process or increase the seed supply. In a critical year for disease development, extensive data can be obtained by having test plots at various locations. Then, too, by testing lines at numerous locations, unusual characteristics of any particular line show up quicker than when only one location is used. Use of multiple test sites decreases the number of years a line must be evaluated.

Release to Growers

The final step in breeding a disease-resistant cultivar is releasing it to growers. First of all, the supply of seed must be increased so enough is available for purchase by farmers. Great care must be taken to keep the seed stocks pure, and the cultivar must be continually tested to make sure disease resistance is maintained. There are several ways to furnish seed of a new cultivar to growers; the procedure varies somewhat in various states and areas. Usually the seed of the new cultivars is produced and sold by certified seed growers as discussed in the remainder of this chapter.

Certified Seed and Seed Storage

The use of poor seed is one of the most costly mistakes a farmer can make. It is impossible to get more out of a crop than the potential of the seed used. Therefore, high-quality seed is a must. No farmer should try to save money by buying cut-rate or cheap seed of unknown quality, as seed costs average less than 3% of the total crop production expenses.

Down through the centuries most farmers have realized the necessity of saving seed for the next crop and improving the seed they planted for the present crop. Two thousands years before Christ, the ancient Hebrews knew the value of pure seed. In Leviticus 19:19, Moses told the children of Israel, "Thou shalt not sow thy fields with mingled seed."

However, before 1900 many seedsmen and farmers had only a hazy concept of varietal purity and its importance. They produced and released supposedly new cultivars with little information about their adaptation, performance, or yield potential. Another serious problem faced by agronomy research workers at this time was the difficulty in obtaining supplies of pure seed of new cultivars produced in other states and countries. Often a new cultivar became so contaminated 3–4 years after release that its identity and value were lost to the seed-buying public.

About 1920, the International Crop Improvement Association, composed of members from the United States and Canada, was formed to get closer and more uniform cooperation between states, coordinate standards, minimize cultivar confusion, and permit sale of certified seed beyond state boundaries. From this beginning, crop improvement associations were established in every state. Usually these associations are designated by state law as the official agency for seed certification. In each state they are associated with the state university or department of agriculture. These associations establish and administer standards for certification of seed and inspect the production of certified seed under these standards (Fig. 6.7). Closely connected with the

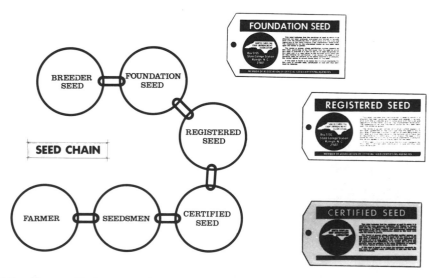

Fig. 6.7. **The certified seed chain.**
Courtesy North Carolina Crop Improvement Assn.

crop improvement association is a foundation seed producers corporation. Both organizations are nonprofit, self-supporting, open organizations composed of seedsmen and farmers or any person or firm who desires to make seed growing a special branch of their farming operation. Membership in these organizations is voluntary.

What Is Certified Seed?

Certified seed is the best seed that is available to farmers. It is high-quality seed that, when planted under normal conditions, will give the desired stand of vigorous, relatively disease-free plants of the intended cultivar. It will not contaminate the land with weeds and other crop mixtures.

Certified seed is of known origin, genetic identity, and purity; it has a known percentage germination. Most of the time it is free of noxious weed seed; often it is disease or insect resistant; when feasible, it has been chemically treated for protection against disease agents.

The basic purpose of the production and sale of certified seed is to make genetically pure crop seed available to farmers. Genetic purity (trueness to cultivar) is established and maintained by special purification and seed-increase programs, by field and seed inspections, and by pedigree records.

The familiar blue label that identifies certified seed assures the user that the seed has passed unbiased field and laboratory inspection processes. This reliable product is in high demand and easily available in seed stores in agricultural communities throughout the United States. Certified seed is a big asset to farmers.

Storing Seed Properly

Because seed is produced for planting at a later date, it must be stored. If stored improperly, seed will deteriorate. Some seeds will die, others will lose their vigor. They will germinate slowly and/or produce weak seedlings. Some seeds will actually rot if stored in moist environments.

Seed moisture is the most important factor affecting the storage life of seed. Each 1% decrease in seed moisture content nearly doubles the safe storage period of seed. It is important to put dry seed into storage and keep it dry. Low temperatures are also desirable. Each 10-degree decrease in temperature will roughly double the safe storage period of seed. Ideal storage conditions are low temperature and low RH, so the seed is kept cool and dry. However, if both moisture and temperature cannot be controlled, then it is more important to control moisture. A rule of thumb for safe seed storage is that the sum of the temperature and RH should not exceed 100 units. For example, if seed is stored at 20°C, the RH should be kept below 80%.

Seed quality cannot be improved through storage, regardless of how good the storage facilities are. Most costly storage problems actually start in the field and result from improper harvesting and processing of seed. Excessive delays in harvest, mechanical injury to seed, and delays in, or inadequate, drying followed by improper storage will result in weak seeds that germinate slowly and produce weak seedlings easily attacked by disease-causing organisms. Excessively high temperature should be avoided in drying seed to be used for planting.

The kind of seed to be stored, the time of storage, and the seed quality level to be maintained all must be considered in planning storage. Different kinds of seed have different lengths of storage life. Peanuts, soybeans, and some other legumes are relatively short-lived in storage; they are not normally carried over from one planting season to the next. However, other legumes (clover and alfalfa), grasses (corn, sorghum, and related species), barley, rye, wheat, and certain vegetable and flower seed can be carried from one planting season to the next under proper storage conditions.

Seed that is carried over during the summer months in general storage needs protection from insects. A variety of insecticides can be used. A local county extension agent can provide a list of recommended chemicals.

Rules for Storing Seed

- Harvest seed when mature. Do not injure it. Dry promptly at proper temperature.
- Store seed in a cool, dry place.
- Do not place seed on the floor or where it will get wet or absorb moisture.
- Protect seed from insects and rodents.
- Run a test to determine germination rate and percentage of any seed of doubtful quality.

Summary

Many successful farmers, gardeners, and landscapers now use the six principal methods of disease management as valuable components of efficient IPM. Disease management practices are interrelated and the use of several technologies in combination is usually more successful and efficient than any one method used alone.

The major types of plant disease management systems are outlined in Fig. 6.8. Some commonly used specific practices are indicated for each type of method. As many crops and ornamental plants are not resistant to all diseases, disease prevention practices are frequently used. The success of these disease management methods depends on the accurate, rapid identification of the disease; the selection of practical and economical management methods; and the proper application and timing of the methods selected.

Plant Disease Management Methods

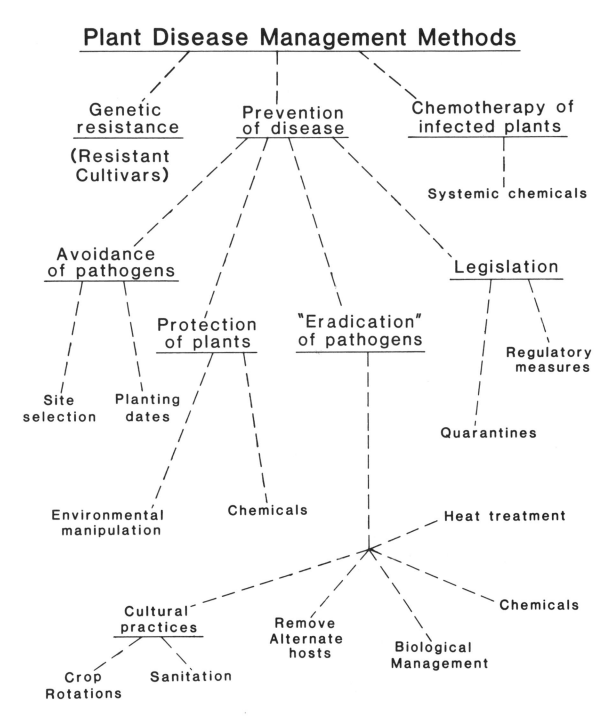

Fig. 6.8. Methods of plant disease management.

7

Chemical Management of Plant Diseases

Some people think that chemical management of plant diseases was invented just a few years ago. But, would you believe that 2000 years ago Homer, the Greek poet, author of *The Illiad* and *The Odyssey,* wrote about the "pest averting sulfur with its property of divine and purifying fumigation." Sulfur is still used today as a pesticide.

Chemical management of plant pests has been an increasingly important method during the last century. The use of chemicals was brought to the attention of the public in 1962 with publication of Rachel Carson's *Silent Spring,* in which the hazards of chemical pesticides were emphasized. Most agricultural scientists agree that pesticides can be hazardous, especially when used by ignorant or careless applicators; however, chemicals are essential in agriculture today. Without them our food supply would be threatened, and it would be impossible to grow certain crops profitably. Moreover, chemical application is sometimes the only economical means for managing plant disease.

Many plant scientists are working diligently to accumulate information to make chemical pesticides safer and yet more effective. Agribusiness is a technical, skillfully organized industry devoted to providing food for an exploding world population, dedicated to the safe and effective use of pesticides. Each day researchers learn more about chemicals and their mode of action. New and safer chemicals are being developed, and it appears that pesticides will be used widely until effective substitutes are discovered.

The use of chemicals for plant disease management is merely one tactical weapon in the battle against plant pests and it should be used in conjunction with the other components of IPM. When used improperly, chemicals are dangerous and expensive, and often will not reduce pest losses. The biggest problem with agricultural chemicals is the people who apply them. Two out of every three pesticide applicators make significant application errors because of inaccurate calibration, incorrect mixing, worn equipment, and failure to read the product label. In 1980, these mistakes, resulting in both over- and underapplication, cost farmers $4–$24/ha in added chemical expense, potential crop damage, and weed competition. In the United States alone, these blundering techniques could cost farmers a billion dollars annually. The increased possibilities of harm to the crop, applicators, farm workers, and consumers are self evident.

The concept that chemical pesticides alone will manage pests is no longer tenable, if it ever was. The application of pesticides to large acreages with little or no regard for their deleterious side effects or impact on nontarget species can no longer be tolerated. We must learn to manage all organisms (including man) in the environment.

How Pesticides Work

Certain chemicals are useful as pesticides because they are more toxic to plant pathogens than they are to plants. Such chemicals inhibit germination, growth, and multiplication of the pathogen, and/or kill it. Some pesticides are *nonselective*, that is, toxic to many pathogens; others are *selective*, affecting only a few or one kind of pathogen.

There is nothing magical about the use of pesticides. To give effective results, a disease must be correctly identified and the correct chemical selected and applied (1) at the correct *time*, (2) in correct *amounts*, and (3) in the correct *way*.

Most chemicals applied on plants can only protect them from subsequent infection and cannot cure a disease after it has started. Also, most chemicals protect only that part of the plant to which they are applied. They are not absorbed into or translocated through the plant but form a chemical barrier between the plant and the pathogen. Thus, when the plant makes additional growth and puts out new leaves, or rain washes off the chemicals, another application must be made.

There are a few chemicals in use today that do have *therapeutic* (*eradicant*) or *curative* properties. These chemicals are called *systemic pesticides* because they are absorbed and spread internally in the sap stream throughout the plant. The oxathins, the benzimidazoles (benomyl, thiabendazole, thiophanate ethyl), the pyrimidines, chloroneb, and ethazol are examples of systemic fungicides. Many others are in the experimental stage and will be available in the near future.

An *antibiotic* is a substance produced by one microorganism and toxic to another. Most antibiotics used today in managing plant diseases are absorbed and translocated systemically by the plant. Streptomycin, a well-known antibiotic, is used against several bacterial pathogens that cause blights, rots, spots, or wilts. Tetracyclines are also active against many bacteria and mycoplasmas. Cycloheximide is effective against several fungi. It is used for the control of turf diseases and is effective against powdery mildews. However, cycloheximide injures some plants, so its use is limited.

Growers should keep up with the latest developments in pesticides and use new chemicals wherever feasible.

Names of Pest Management Chemicals

There are literally hundreds of chemicals in use today for crop protection. They are applied as aerosols, dusts, fogs, granules, mists, paints, pastes, smoke, soaks, and sprays. New ones constantly are being developed. Therefore it is necessary to name them properly and accurately. To avoid confusion and mix-ups, a committee composed of representatives from professional societies and industrial organizations was established in 1954 to prepare detailed specifications and procedures for naming pesticide chemicals. At the present, every pesticide has three names:

1. *Generic, common,* or *coined name* for all compounds having the same active ingredient
2. *Chemical name*
3. *Trade name* (brand or manufacturer's name)

For example, the common name for the chemical tetrachloroisophthalonitrile is chlorothalonil. Trade names for it are Bravo and Daconil 2787. Each pesticide differing in the ingredient statement, analysis, manufacturer or distributor, name, number, or trademark is considered as a distinct and separate brand.

Pesticide Safety

Pesticides are beneficial when properly used. They may be extremely dangerous when used improperly. Most pesticides are designed to kill something—animals, bacteria, fungi, insects, mites, nematodes, or weeds. Therefore, all pesticides should be handled carefully and all safety precautions observed (Fig. 7.1A). Gasoline and table salt, when used in the wrong way, are also dangerous and poisonous. Nonetheless, careless or ignorant persons should not handle pesticides.

Goggles

Respirator

Long-sleeved shirt

Sleeves over long rubber gloves

Wide-brimmed hat

Overalls

Rubber band around cuff

Rubber boots

Fig. 7.1A. **Protective clothing to be worn when applying hazardous pesticides.**
Courtesy J. H. Wilson

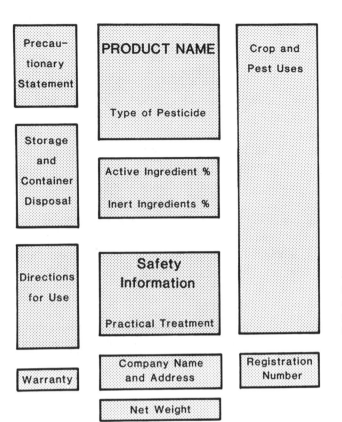

Boxes in figure:

Precautionary Statement	PRODUCT NAME Type of Pesticide	Crop and Pest Uses
Storage and Container Disposal	Active Ingredient % Inert Ingredients %	
Directions for Use	Safety Information Practical Treatment	
Warranty	Company Name and Address	Registration Number
	Net Weight	

Fig. 7.1B. The label attached to a pesticide container describes how to use the product correctly and what special safety measures should be taken.
Courtesy USDA

Read the Label

The pesticide label contains the most important information you can learn about a particular pesticide. Therefore, before buying, mixing, applying, storing, or disposing of a pesticide, *read the label.* This means you must read the label at least five times. The information found on a pesticide label is outlined in Fig. 7.1B.

Pesticide labels are required by law to carry certain signal words, such as *Danger* (skull and crossbones), *Poison, Warning, Caution,* and *Keep Out of the Reach of Children,* depending on the hazard of the particular product. Labels with skull and crossbones on them are highly hazardous, whereas Caution on a pesticide label indicates a relatively nonhazardous chemical. Table 7.1 gives the meaning of the signal words.

Table 7.1. Signal Words and Meanings on Pesticide Labels

Signal word[a]	Meaning	Lethal dose[b]
Danger Poison (in red)	Highly hazardous	Taste to a teaspoon
Warning	Moderately hazardous	Teaspoon to a tablespoon
Caution	Slightly hazardous	30–More than 500 ml

[a]The statement "Keep Out of the Reach of Children" is required on all pesticides.
[b]For humans when taken by mouth (orally).

Checklist of Safety Precautions

The following basic rules should be observed by all persons who use pesticides:

- Be careful. All chemicals are dangerous when handled carelessly.
- Read the label on the container.
- Use pesticides only when needed.
- Do not eat, drink, or smoke when applying pesticides. Keep pesticides away from food and drinking water.
- Know what to do in case of an accident. Post physician's phone number by phone.
- If pesticides are spilled on the skin, wash off immediately with water. Take off contaminated clothing and wash it before wearing again.
- Take time to explain the safe use of pesticides to employees.
- Check application equipment for leaks, clogged lines, nozzles, and strainers.
- Calibrate equipment frequently.
- Cans, bottles, and drums that have contained pesticides should not be used for any other purpose. Handle and discard containers in a recommended manner.

Pesticide Residues and Tolerances

Modern analytical methods make it possible to detect very small amounts of pesticides in plant or animal products. Pesticides used on young plants, for example, may be found on the harvested portion of the crop. Quantities as small as one part per million (ppm) and even one part per billion (ppb) can be detected on crops. One part per million (ppm) is 1 mg in 1 kg, or 1 lb in 500 tons, or 1 oz in 31½ tons, or 1 oz in 63,000 lb.

What Is a Residue?

A *residue* is the amount of a pesticide that *can be found on, or in, a harvested crop* (or slaughtered animal). Residues can come from pesticides applied directly to the crop, from a nearby crop by drift, from uptake from the soil, and even from storage in an area previously treated with a pesticide.

Some pesticides are very persistent and last for a long period of time, even years, after application; other pesticides are not as persistent. Naturally, the persistent pesticides are more likely to cause long lasting residues.

Pesticide applicators must do everything they can to avoid illegal residues. Many pesticides have warning statements on the label such as *Do Not Apply Later Than 7 Days Before Harvest,* or *Do Not Apply Within 30 Days of Slaughter.* These directions are given to prevent residue problems on some crops. If label directions are followed and residue problems develop, the manufacturer is responsible; if the label is not followed, the applicator or grower is responsible.

Most pesticides decompose in the soil, leaving residues that are not toxic to plants nor taken up by plants in sufficient amounts to be harmful. An exception is bromine, which is a part of dibromochloropropane (DBCP) and ethylene dibromide (EDB). Bromine is toxic to garlic, onions, and a few other plant species. It is readily taken up by other plants and is detectable in milk from cows fed on peanut hay from fields where DBCP or EDB has been used.

What Is a Tolerance?

A *tolerance* is the amount of pesticide residue that by law may remain on or in food. The tolerance for a pesticide is set by the Environmental Protection Agency and represents an amount of the pesticide that will not cause injury to consumers.

A tolerance may be set for each pesticide and also for each specific use of a pesticide. For example, the tolerance for Pesticide-X on beans might be 2 ppm and for alfalfa it might be 1 ppm. If a tolerance is exceeded, the crop or animal product can not be sold and the grower may be liable.

Pesticide Formulations

Once the proper pesticide has been selected, it must be applied to the target. Many factors affect our ability to place the pesticide on the target in the manner and amount for most effective results, with the least undesirable side effects, and at the lowest possible cost. The selection and use of equipment is of utmost importance. In modern agriculture elaborate and expensive application equipment sometimes is necessary. Moreover, without proper consideration to formulation, compatibility, and use of records, successful application is impossible.

Pesticides are usually not used in their pure form. The *active ingredient* is usually mixed with inert substances. This mixture is called a *pesticide formulation*. A pesticide formulation must be prepared so the user can apply it in a safe, convenient, and effective manner with available or easily modified equipment. Available pesticide formulations include dusts, emulsifiable concentrates, flowables, granules, soluble powders, solutions, and wettable powders. Combinations of pesticides are sometimes used, for reasons of economy, when management of more than one pest is required. Adjuvants are added to some pesticides to increase their effectiveness.

Dusts (D)

A dust formulation consists of a pesticide mixed with finely ground talc, clay, pyrophyllite, powdered nut hulls, or other such inert materials. Dusts must be uniform in particle size.

Dusts are used dry and should not be mixed with water. Dust formulations are used to treat seed and growing plants to manage insects, diseases, or other pests. Dust formulations require special equipment for application, may be wasteful, and may be more hazardous to use than other pesticide formulations.

One drawback to the use of dusts is that wind, mechanical rubbing, and rain all remove dust from plant surfaces. A film of water on plant surfaces increases the sticking power of dusts. Thus, it is more efficient to dust at night when leaves are wet with dew or when there is no wind, if this is possible. Dusts may have to be reapplied at frequent intervals to provide adequate protection.

Emulsifiable Concentrates (EC)

An emulsifiable concentrate is a liquid formulation of a pesticide that forms an *emulsion* when mixed with water. An emulsion is one liquid dispersed throughout another type of liquid. Many active ingredients in pesticides are not soluble in water but are soluble in other liquids; in an EC the active ingredient may be dissolved in an oil and an emulsifying agent added. The EC can be mixed with water to form a "milky" emulsion. Emulsifiable concentrates are not abrasive and are relatively harmless to equipment. Some plants are sensitive to some of the solvents or emulsifiers used in these formulations. An EC must be continually agitated during application to prevent settling out of the active ingredient.

Flowables (F)

A flowable formulation is a finely ground wettable powder that is sold as a thick suspension in a liquid.

Granules (G)

A granular pesticide formulation is usually prepared by *impregnating* (soaking up) a liquid pesticide to granules made from clay, corn cobs, walnut shells, or other porous materials. Granules are prepared to a standard size that are described by mesh size. For example, all of the particles in a 20/40 mesh granule will pass through a screen with 20 openings per inch, but none will pass through a screen with 40 openings per inch.

Granular pesticides are often applied to the soil. Since they do not cling to plant foliage, they can be applied over plants and fall to the soil. Application of a pesticide formulated as granules is usually safer than liquid or dust formulations of the same pesticide.

Soluble Powders (SP)

Soluble powders are powdered materials that will dissolve in water.

Solutions (S)

Pesticides formulated as solutions usually contain the pesticide and a solvent. The solvent may be a petroleum product or water. The type of solvent used to prepare the solution will depend upon the nature of the pesticide and the use of the formulation. Solutions prepared for special uses should not leave unsightly residues. Many of these preparations with petroleum products will damage living plants if not used properly.

Wettable Powders (WP)

Powdered pesticide formulations that contain wetting agents are called wettable powders. The active ingredients are usually more concentrated than in dusts. Wettable powders form a suspension in water with agitation that should not settle out too quickly. Wettable powders are less likely than emulsifiable concentrates to damage sensitive plants, but are more abrasive to pumps and nozzles.

Combinations of Pesticides

Plant diseases and insects often attack a plant at the same time, and sometimes growers may wish to protect crops against several diseases. Therefore, for reasons of economy the use of multiple combinations of pesticides is a necessary practice used by many growers in commercial agriculture. The utilization of multiple pesticide combinations, often called *tank mix, piggyback,* or *one-shot* applications, is a laborsaving shortcut. Two fungicides may be combined, or a fungicide and an insecticide, or a fungicide, and growth regulator plus minor elements. However, only compatible pesticides can be used in combinations or mixtures.

Compatible chemicals are those that can be mixed without affecting the effectiveness, physical properties, or phytotoxicity of either chemical. Incompatible chemicals are those that, when mixed, lose their effectiveness or injure the plant. They should not be mixed. *Physical incompatibility* often results when "hard" water is used for spray mixtures. In such cases the pesticide tends to fall out of suspension. Pesticides also may be *chemically incompatible,* that is, they react together so as to lose the effectiveness of one or all components. For example, most organic pesticides should not be mixed with alkaline compounds with a pH higher than 7.0. Chemical incompatibility frequently is the cause of poor performance of some multiple pesticide combinations. When two or more chemicals are mixed and applied at the same time, each must end up in the correct place (*placement compatibility*) if they are to be efficient. Some fungicides must be uniformly distributed over the plant to get efficient disease management, whereas certain insecticides need to be directed only at the top of the plant; mixtures of such chemicals would lack placement compatibility. Before combining pesticides, read the directions on the label and avoid the risk of plant injury problems. Pesticides should be mixed only if a particular combination is specified on the label. Avoid combinations of pesticides of unknown compatibility.

Adjuvants (Additives)

Many plants have leaves with a protective waxy coating, which makes them slick and smooth. Others have leaves covered with hairs called fuzz. Smooth cabbage or onion leaves, young apples, or fuzzy peaches are difficult to cover properly and uniformly with sprays. Or the spray may "ball up" and run off, leaving unprotected areas on the leaves or fruit. Therefore some chemicals require an *adjuvant,* which is a material added to a formulation to improve its physical or chemical characteristics. Adjuvants are commonly used to improve the spreading ability of a pesticide or the "weathering" of the pesticide deposit after application.

Adjuvants include detergents, dispersing agents, emulsifiers, foam suppressants, penetrants, safeners, soaps, spreaders, stickers, and wetting agents. Most pesticides already have the necessary adjuvants in the formulation. Adjuvants may have a negative effect if not needed. Therefore, they should be used only when called for by the pesticide label or when recommended by reliable research data.

Soil Treatment

A big step forward in the management of soilborne pathogens and root diseases occurred about 1950 with the development of *fumigation* techniques to inject chemicals into the soil to control nematodes, fungi, and bacteria. Today soil treatment with chemicals is a necessary component of IPM for many crops. Some crops cannot be grown profitably in specific areas unless the soil is first treated with chemicals. For instance, in 1979 almost 90% of the acreage planted to tobacco in the southeastern United States was treated with soil chemicals prior to transplanting. Many fields planted to crops that return a high net income per hectare (for example, cotton, ornamentals, strawberries and vegetables) are routinely treated before use. Flower beds, greenhouse soils, golf greens, and lawns often are treated before the seed is sown or the crop is planted. Fumigation pays off for tree crops, too. Preplant fumigation is recommended as a standard practice before planting apple, banana, citrus, or peach trees. The manufacture, sale, and application of soil-applied plant pesticides is now estimated to be a billion-dollar-a-year industry. Undoubtedly additional chemicals will be developed in the near future that will give more efficient control of soilborne pathogens and pests, thus helping to increase yields and maintain our food supply.

Fumigants

There are two types of soil-applied pesticides—*fumigant* and *nonfumigant.* A fumigant produces a gas, vapor, or fume in the soil that kills not only nematodes but also many bacteria, fungi, insects, rodents, and weeds. Some soil fumigants are applied directly to the soil; others are applied under gas-proof covers made of polyethylene or similar materials. Fumigants may be used to control a single pest or a number of pests. The latter, called *multipurpose fumigants,* are most useful in fields where the soil is infested with more than one pathogen.

All fumigant pesticides are applied as gases or have a gas phase after they are applied. The gases diffuse through the air in the soil, are dissolved in soil water, and enter the bodies of nematodes through their skin. Similarly, these toxic chemicals are absorbed through the cell walls of fungi and bacteria and kill them.

Factors Affecting Fumigant Performance

One big disadvantage of some fumigants is that they are phytotoxic. Crops planted in soil containing some fumigant nematicides may be injured or killed. Therefore, many fumigants must be applied as *preplant* treatments after which a waiting period of 10–21 days is required between fumigant application and planting date. This permits the poisonous fumes to evaporate and diffuse out of the soil. Several factors influence fumigant performance and the required waiting period between application and planting.

Temperature. Fumigants, once they are applied, must become gases to be effective. If the temperature is high, the conversion to a gas is rapid; if it is low, the conversion is slow. However, if the temperature is too high or too low, the conversion may be too rapid or too slow to give a proper kill. Check the label to determine the proper temperature range for applying a particular fumigant (10°–25°C for most fumigants).

Moisture. Fumigants that are applied in the soil or to fumigate grain are affected by the moisture content of the soil or grain. Fumigants applied in soil with a moisture content that is too high or too low may not give efficient pest management. In dry soils, fumigants may escape too rapidly; in wet soils, water occupies most of the air space so the fumigant moves very slowly. Soil fumigants perform best under average soil moisture conditions even though some move in the liquid phase.

Soil Condition and Type. Fumigants move faster in loose and sandy soils than in clay or compact soils, other factors (temperature and moisture) being comparable. Fumigants applied to a cloddy or poorly prepared soil are likely to be ineffective. Highly organic soils and soils with plant residues (stalks, vines, leaves) may "tie up" the fumigant. Fumigant rates may have to be increased in such soils. Proper planning and soil preparation can reduce this problem.

Timing and Aeration. Proper timing is important in applying most fumigants. For example, with seedbeds the temperature is usually higher and fumigants perform more efficiently in the fall than in the spring. The exposure time, or time required to kill pathogens, varies with the fumigant applied, temperature, soil type, crop, and other factors.

Aeration time, or the time required to allow the fumigant to leave the soil before planting, is important to avoid injury to the crop. Proper aeration is also necessary to assure the adequate dispersal of the fumigant before workers enter a fumigated building, grain bin, greenhouse, seedbed, or other area.

Other Factors. Some crops are very sensitive to certain fumigants. Even small quantities of a fumigant or fumigant residues can harm some crops. *Read the label* to find out if the fumigant can be used before a crop is planted or on living plants and the limitations, if any, on the crop in question.

Broad-spectrum fumigants applied under covers can temporarily harm nitrifying bacteria essential for converting ammonium to nitrate nitrogen. Nitrate fertilization may be required to correct this condition.

Applying Fumigants

The skill and accuracy with which pesticides are applied is just as important as selecting the correct pesticide. For effective use of soil fumigants, careful attention to application details must be followed.

Before fumigation, the soil should be prepared in the same manner as for planting a crop; that is, the soil should be moderately loose and free of undecayed roots or other partially decomposed crop residues. Fumigants perform best in fairly moist soil; results may be unsatisfactory when fumigants are applied to dry soil. Moderate rain following application is beneficial. Heavy rains soon after application chill the soil and delay escape of the fumes; hence the waiting period before planting must be extended. The soil temperature should be above 5°C, preferably between 15° and 27°C. Fumigants should not be applied on windy days because the evaporation rate is too high.

In today's modern agriculture there are many types of equipment for applying farm chemicals, ranging from hand applicators or one-row equipment for the homeowner or small farmer to sophisticated multirow tractor-drawn applicators capable of treating many hectares in an 8-hr day. All farmers should learn the proper care, maintenance, calibration, and use of equipment used to apply farm chemicals.

Hand Applicators. For crops planted in widely spaced rows (e.g., watermelons) or when only a few square meters of soil are to be treated, a large saving of fumigant is possible by use of "spot" treatments. A single injection of liquid fumigant about 20 cm deep produces a spot of treated soil 30–40 cm in diameter in which one plant can grow. "Spot" or "site" treatments can also be used for perennials (e.g., apple trees). Treated sites range in size from 1 to 3 m in diameter according to the size of the tree, bush, or vine to be planted. Hand applicators are commonly used for spot treatments with liquid fumigant. Liquid from a 1-liter tank is measured by a pump with an adjustable stroke and delivered beneath the soil surface through holes near the point of the soil spike.

Low-Volatile Liquid Fumigators. *Low-volatile* fumigants are handled in open or nonpressurized containers. The fumigators for these fumigants may be gravity or forced-pressure fed and are designed for row or *broadcast* applications. Most row fumigators are the constant-head gravity type, whereas broadcast fumigators are force-fed types. The forced-fed type requires no agitation, and the liquid is delivered into the soil through an injector. Positive-displacement piston-type applicators are also used for broadcast application, but their accuracy is greatly reduced at high speeds. The more volatile fumigants may vaporize in the suction lines and pump on the intake stroke causing the pump to pump vapor instead of liquid.

Chisel-injection method. In the chisel-injection method, used for broadcast (overall) treatments, the liquid fumigant is applied in bands not more

Fig. 7.2. A broadcast applicator for liquid fumigant nematicides. Nematicide from the tank flows through a regulating apparatus to each of the six curved shanks and is injected at a depth of about 20 cm. Sometimes a drag is used behind the rig to firm and seal the soil.
Courtesy A. L. Taylor and J. N. Sasser

than 30 cm apart and 20–25 cm deep (Fig. 7.2). These large-scale applications are made with power-driven applicators of the continuous flow type that deliver from 5 to 10 streams into the soil simultaneously. Applicators consist of a tank to hold the fumigant and a system of tubes to convey it into the soil, the tubes passing down the back edge of narrow cultivator teeth or "chisels." One or more pumps are included in the mechanism, and the liquid is delivered under pressure. The chisels are mounted on a horizontal bar or framework, which can be raised or lowered. The applicator may be mounted directly on a tractor or carried on a trailer. A drag is pulled behind the applicator to firm the top layer of soil, thus sealing in the gas. Irrigation water is often needed when low-volatile fumigants are applied to growing plants such as turf or orchards.

Row method. The row method of application, in which only the row area is fumigated, requires about half the amount of fumigant as does the chisel method. Usually one or two streams of fumigant nematicide or a band of nonfumigant nematicide about 25–30 cm wide is applied for each row. The area between rows is left untreated. The crop is planted in the treated area and by the time the roots have grown into the nontreated area, the plants are large enough to escape serious damage. Many vegetable and field crops are treated by the row method.

Row treatment is less expensive than other types of treatment, but has several disadvantages:

- It can be used most efficiently only as a spring treatment.
- It must be carefully timed and fumigant applied at least 2 weeks before the recommended planting date for the locality. Bad weather may upset these calculations.
- When heavy rains occur during the waiting period before planting, the fertilizer applied with the fumigant may leach away.
- Since only the row area of the soil is fumigated, the pest population may build up faster than with broadcast treatments.

High-Volatile (Gas-Type) Fumigators. *High-volatile* fumigants, such as methyl bromide, must be handled in closed pressurized containers. The fumigators are designed for force feeding these fumigants from a pressurized tank. The tank is either precharged with enough pressure to empty its contents or pressurized gas is fed into the tank during application to displace the fumigant.

The precharged tank cylinders are approximately two-thirds filled in order to have room for the precharging gas. The fumigant is injected into the soil and covered with plastic to prevent its escaping. While these applicators are referred to as "gas type," the fumigant must remain as a liquid in the tank, pressure lines, manifold, and metering devices for accurate application.

Treating seedbeds. High-volatile fumigation is often used for treating seed beds. The general procedure is as follows:

1. Air temperature should be at least 13°C, preferably higher. However, it should not be above 32°C as the plastic cover begins to leak gas if exposed to continuous high temperatures. When daytime temperatures reach 32°C, treat the beds late in the afternoon to prevent excessive heat under the cover. Soil should not be too wet; it should be in good working condition.

2. Prepare the seedbeds as if for planting. Crop material should be well decomposed. Sticks, dead stems, and sharp rocks will puncture the cover and let the fumes escape prematurely. Work in fertilizer if bed is to be sown immediately after treatment.

3. Cover the bed with a plastic cover. Support the cover to about a height of 10 cm with bags, small mounds of soil, or other suitable objects; seal three edges with soil, then flap the open end to force air under the cover.

4. Release 9 cans (454 g each) of methyl bromide per 100 m² of bed, with the aid of an applicator, into open containers to catch methyl bromide liquid until it evaporates.

5. Leave cover in place 24–48 hr. Sealing fumes into the soil for more than 48 hr may cause injury from bromine absorbed on soil particles. Germination may be reduced and/or delayed and plants may be stunted and retarded. Aerate soil by removing the cover.

Many farmers who must fumigate several hectares of seedbeds prefer to hire commercial applicators to treat the soil, using large rigs that inject the fumigant and lay the cover simultaneously (Fig. 7.3). However, the same rules

Fig. 7.3. Fumigating a seedbed with methyl bromide. After the soil is prepared this tractor-drawn chisel applicator applies the methyl bromide, lays and seals the polyethylene cover all in one operation.
Courtesy Reddick Equipment Co.

apply. The soil must be prepared properly and the correct fumigant applied at the correct time, in correct amounts, in the correct way.

Nonfumigants

Another type of soil treatment chemical was introduced about 1960—the *nonfumigant* nematicides, some of which also kill insects. Nonfumigant nematicides dissolve in the soil water and are distributed through the soil. Some of these nematicides are systemic in plants. These are absorbed by the plants through the roots after application to the soil; or they can be applied to the leaves and absorbed through them, then translocated to kill nematodes feeding on the roots. Presumably they are taken up by the nematode during feeding, but they may also enter through the nematode's skin (cuticle) in contact with plant tissue. Nonsystemic nematicides remain in the soil water and enter through the cuticle of the nematodes. A big advantage of non-fumigant nematicides is that there is no waiting period: the crop can be planted immediately after the nematicide is applied.

Granular Application

Nematicides in granular form are applied to gardens, pastures, row crops, and turf. Depending on the chemical, granules may be used because they are safer to handle (less hazardous to applicators), more convenient to apply, less phytotoxic, and/or more selective in their action than other formulations. In some cases, granules are preferred because they permit more effective placement of the chemical. As with other pesticide formulations, however, granules must be handled carefully because most are toxic to humans.

Granular application equipment includes small hand-operated applicators and large truck- or tractor-mounted or tractor-towed units. Granules may be applied in a band or broadcast uniformly (Figs. 7.4–7.6). The granules are distributed after they are fed from a hopper by deflecting baffles for band or broadcast application, or by a centrifugal thrower for broadcast spreading. They are either incorporated into the top 10–20 cm of soil by disking or spread on top of the soil (or turf) and watered into the soil.

Granules are seldom used for foliar application because they will not stick to the leaves. One advantage of using granules is that no mixing is required. However, they generally cost more than other formulations.

A granular applicator should be easy to clean and fill. It should have mechanical agitation over the outlet holes to prevent bridging and to keep the flow rate constant. Application should stop when drive stops even if the outlets are still open.

Liquid Application

Liquid nonfumigant nematicides may be diluted and sprayed uniformly over the soil surface and then disked into the soil. For treating gardens or small areas, hand sprayers can be used; for large fields, low-pressure (boom-type) tractor sprayers are often used (see Fig. 7.8).

Seed Treatment

Many pathogens are carried on or in the seed, and seeds are often planted in soil infested with pests. To protect the seed and the developing seedling from attack, various protectants can be applied as dusts or sprays; or the seed can be soaked in a pesticide solution and then allowed to dry. Tubers, bulbs, and roots also may be chemically treated to protect them from diseases or insect pests. Usually a bright-colored red, blue, or yellow dye is included in the chemical mix to stain the seeds, indicating that they have been treated. Seed treatment is a specialized business and is best performed by skilled operators. Therefore, this operation is usually conducted by seed-distributing agencies rather than by individual growers. Some fungicides vary in their ability to manage certain pathogens. Since some chemicals injure certain kinds of seed and not others, specialized equipment is required to properly treat the seed. The cost of treated seed is usually so small (only a few cents per hectare) in relation to the value of the crop that few farmers can afford *not* to use this form of insurance.

Caution: Chemically treated seed should *never* be used for livestock feed or food for human consumption. Sacks, bags, or other containers should be thoroughly cleaned before reuse or discarded. Do not handle treated seed with bare hands.

Hopper

Feeder Gate
Control Dial

Feeder Tube
From Hopper

Band Applicator

Fig. 7.4. Diagram of row-type applicator used to apply granules at planting time.
Courtesy Am. Assoc. Vocational Instructional Materials

Fig. 7.5. Four-row planter applying seed, fertilizer, and pesticide granules all in one operation. Note shields to guard against wind drifting.
Courtesy Gandy Co.

Fig. 7.6. Pull-type broadcast granular applicator, 14.4 m wide. This rig can apply pesticide granules to 120 ha per 10-hr day. The two outside sections can be folded for transport or road travel.
Courtesy Gandy Co.

Postharvest Treatment

Some agricultural products must be stored several days to several months before they are used. During this time they may begin to rot. A number of chemicals can be used to reduce postharvest or storage diseases. Some may be applied just prior to harvest; others, in the wash water during processing, or by using wrapping paper impregnated with fungicidal materials or by periodically fumigating the storage house with volatile chemicals. However, some chemicals leave harmful residues on the stored product.

Residues are a big problem in the use of postharvest chemicals. Before any material can be used, the Food and Drug Administration (FDA) sets a limit on the amount of pesticide residue that can be present on materials intended for consumption. Exhaustive tests must be conducted with chemicals before they can legally be used as postharvest treatments. Nonetheless, many postharvest diseases are successfully controlled by chemicals, and the search for new, safer chemicals continues.

Treatment of Growing Plants

The greatest volume of chemicals used for plant disease prevention are applied as sprays or dusts to leaves, flowers, or fruits. At present, more fungicides are used than bactericides. Most chemicals are protectants and must be present on the plant surface before a fungus spore or bacterium arrives there. Therefore, it is important that the entire plant be covered with the chemical. Growing leaves, branches, and fruit must receive repeated applications to protect the new growth.

Reducing Treatment Costs

The application of pesticides to plants, either as dusts or sprays, can be an expensive, time-consuming, often disagreeable task that must be done carefully if efficient pest management is to be achieved. Many pesticides are derived from petroleum products, and it takes fuel to drive the machines that apply pesticides. With a dwindling supply of fossil fuels, energy conservation is a necessity. It appears inevitable that fuel costs will continue to rise. Therefore, pesticide applicators are always on the lookout for ways to reduce costs.

For certain diseases, spray or dust "schedules" have been developed that tell the farmer at what intervals pesticides should be applied, depending on the disease, frequency and duration of rains, RH, temperature, and time of year. For other diseases (apple scab, late blight of potato, peanut leaf spot), recording devices have been developed that automatically monitor such critical factors as temperature, RH, rain, and wind. These "little gray boxes" are placed in the crop area and collect and process data using a program on a microcomputer; based on such data, a forecaster can tell farmers whether it is necessary to spray. This practice sometimes results in great savings in the costs of chemicals and the expense and time of application.

Another way to reduce treatment costs is to make more dust and spray particles stick to the plant surfaces. Equipment is being developed to charge each particle electrically. With such *electrostatic* dusting or spraying equipment, the negatively charged particles are attracted to the positively charged plant surface, so that more uniform coverage is obtained and less material is required. Battery or motor-driven electrostatic charging units can be attached to most field equipment with slight modifications. Electrostatic application fits in well with *ultra low volume* (ULV) application and may be one of the improvements available in the near future.

Another type of sprayer that is being developed is one that recycles the chemical that does not stick on the plant. Spray nozzles are directed such that the extra spray is caught and returned to the spray tank rather than being wasted on the soil. This type of sprayer will permit more efficient use of chemicals and reduce the hazard of excess chemicals in the environment.

In the early 1950s airblast sprayers came into general use by growers with large acreages. These sprayers had powerful engines that could envelop a tall tree with a fog of pesticide in a few seconds or cover up to 20 rows of cotton in one sweep. At that time fuel and pesticides were cheaper than they now are. At about the same time, lightweight, durable compact pumps driven by the power take-off (PTO) systems of tractors and more durable nozzles came into general use for conventional-type sprayers.

Application costs also have been cut by reducing the amount of inert carrier or water used to dilute pesticides or by eliminating the water altogether. Not only is it expensive to haul water back and forth across a field either in a sprayer or in an airplane, it is often difficult to have clean water convenient for conventional spraying techniques. The application of small quantities of pesticide, as little as 10 liters/ha, is referred to as *ultra low volume* spraying or dusting. Eliminating the diluent lessens the expense of applying pesticides enough that application of pesticides by airplane, helicopter, or multirow dusters or sprayers to low-value crops in large fields (e.g., wheat, corn, and other grains) becomes feasible.

The first concentrate or ULV sprayers were introduced in the early 1960s. These sprayers made it possible to reduce the volume of water used per hectare by breaking up the spray into smaller droplets through the shearing effect of high air velocity. Later, centrifugal energy nozzles led to development of *controlled droplet application* (CDA), in which sprays are applied in a narrow range of droplet sizes (usually less than 100 μm) selected according to the target at which the pesticide is aimed. Droplet size is determined by the rotational speed of a spinning disk inside a hollow cone in the sprayer head.

Controlled droplet application is drastically different from conventional application with the commonly used boom sprayer, which makes droplets by forcing the liquid spray under pressure through various types of nozzles (see Spray Equipment Parts section). The droplets produced by conventional pressure sprayers range in size from 20 μm to 400 μm. Many of the very small droplets drift away or bounce off leaves and are lost. The large droplets are ineffective in covering leaf surfaces so certain areas of the leaves are unprotected. In contrast, CDA produces spray droplets of uniform size that are highly efficient in reaching and covering plant surfaces, even under windy conditions. The uniform droplets give maximum penetration of the plant canopy and cover both upper and lower leaf surfaces. In addition to saving at least 50% of the pesticide costs, CDA results in less chemical residues in the crop.

At the present time, ULV tractor-mounted or aerial equipment is widely used in some areas. Small hand-carried, battery-operated sprayers for the small farmer are also available. Further refinement of ULV application will undoubtedly lead to more efficient chemical management in the future.

In the remainder of this chapter, various types of spraying and dusting equipment, general application guidelines, calibration of application equipment, and environmental considerations related to the use of pesticides are described in more detail. Much of this information applies specifically to treatment of growing plants with sprays or dusts. However, some of the information applies to other types of treatment or to the use of pesticides in general.

Pesticide Sprayers and Dusters

Sprayers are the most commonly used pesticide application equipment. There are many types and sizes ranging from small handheld sprayers for home and garden to large tank models for spraying crops, golf courses, greenhouses, nurseries, orchards, roadways, tall trees, and vineyards. Proper spraying requires skill and accuracy. Moreover it is expensive. Therefore, careful attention should be paid to selecting the proper equipment to do an efficient job. In addition proper care and maintenance of valuable spray equipment is a necessary component of efficient application.

A large sprayer for commercial use is a long-term capital investment. It should be selected carefully based on the following considerations:

- Does the size and design of the sprayer fit its intended use?
- Is the sprayer designed so that it is easy to fill, clean, handle, operate, and adjust?
- What is the quality of material and construction of the tank, pump, agitator, and strainers?
- Are spare parts and repair service quickly available?
- How many people are needed to operate the sprayer at maximum efficiency?

For proper and safe operation, a sprayer must be in good condition in the hands of a well-trained operator. The operator's manual for all spray equipment contains instructions on how to use and care for it. To keep a sprayer at peak performance, it must be serviced and adjusted before each use (see Calibration of Equipment section). Sprayers are subjected to hard use, and they are exposed to corrosive and abrasive chemicals. To prevent problems during future use, a sprayer should be cleaned and stored properly after each use (see Cleaning and Storing Equipment section), and checked and serviced regularly.

Hand Sprayers

Hand sprayers are used for spot treatment and home garden use. Most are displacement-type sprayers; that is, air is pumped into a pressurized tank thereby displacing the spray liquid, forcing it out through the nozzle. Most hand sprayers are economical, simply constructed and easy to use, clean, and store. However, many lack good screens and clog easily. The only way to agitate the spray liquid is by shaking. There are five types of hand-operated sprayers:

1. *Aerosol* sprayers have sealed containers of compressed gas and pesticide. They usually are less than 1 liter in capacity and cannot be reused when empty.

2. *Trigger-pump* sprayers are small, refillable units with a trigger-action device.

3. *Hand-pump* sprayers are lightweight, push-pull, refillable units with a capacity of about 1 liter.

4. *Compressed air* sprayers vary from 4 to 20 liters in capacity. Air is compressed by a piston-type hand pump. Both hand-carried and backpack models are available. Air pressure is not constant and the repeated pumping required to maintain pressure is tiresome and time-consuming. More expensive constant-pressure sprayers are equipped to feed carbon dioxide or compressed air into the spray tank through a regulator valve.

5. *Garden hose* sprayers are operated by pressure from a water system (Fig. 7.7). When the water is turned on, concentrated spray liquid is siphoned out of a container of about 1-liter capacity attached to the water hose. The pesticide is diluted by and mixed with the stream of water as it discharges.

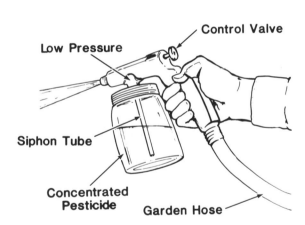

Fig. 7.7. **Sprayer attached to garden hose.**
Courtesy Am. Assoc. Vocational Instructional Materials

Small General Use Sprayers

The *power wheelbarrow sprayer,* the *estate* or small wheel-mounted sprayer, and the small *skid-mounted sprayer* are classified as small general use sprayers. They are designed to handle spraying jobs too large for hand equipment, such as large garden, greenhouse, nursery, and small tree spraying.

Power wheelbarrow sprayers have alternating piston pumps that deliver 6–12 liters/min and develop pressures up to 250 psi (pounds/square inch). Power is supplied by an air-cooled engine of 1.5–3 hp (horsepower). Tank size ranges from 50 to 75 liters. Spray material in the tank may be mechanically or hydraulically agitated. Standard equipment includes an adjustable hand gun and a hose. Spray booms are available for some models.

Estate sprayers are small two-wheel mounted units equipped with handles for moving. Trailer hitches are available as special equipment. Smaller models have 60- to 120-liter tanks and pumps that deliver 6–12 liters/min at pressures up to 250 psi. Larger models have 800-liter tanks and pumps that deliver 12–16 liters/min at pressures up to 400 psi. Power is supplied by a 1.5- to 3-horsepower air-cooled engine. Spray material in the tank is mechanically agitated. Standard equipment includes a hand gun and a hose. Spray booms are available for some models.

General Use or Multipurpose Sprayers

General use sprayers are designed to handle most of the spraying needs on general farms from low-pressure weed or row-crop spraying (30–40 psi) up to medium-pressure field-crop and livestock spraying (60–80 psi).

These sprayers may be tractor mounted, pull type, or self-propelled (Fig. 7.8). They usually have roller-type pumps that limit their pressure to about 80

Fig. 7.8. Boom-type, tractor-mounted, general-purpose sprayers can be adapted to apply many different kinds of pesticides on the average farm.
Courtesy W. A. Skroch

Fig. 7.9. An airblast sprayer may cover a swath up to 27 m wide and may spray up to 16 ha/hr. A high-speed, fan-driven air stream breaks the spray material into fine droplets as it comes from the nozzles. Sprays may be directed to one or both sides and higher or lower, as desired.
Courtesy F. E. Myers and Bros. Co.

psi for continuous spraying. Higher pressures cause very rapid pump wear. Centrifugal pumps are available for low-pressure spraying when larger volumes are needed. Piston pumps are available for spraying of higher pressures. Tank sizes range from 200 to 800 liters. The spray material is usually agitated hydraulically. Agitation is often a problem with larger tanks using a roller or piston pump.

This type of sprayer is used to apply a wide range of pesticides and fertilizers. Good-quality materials for the tank, pump, and other components are essential to reduce corrosion problems. A general use sprayer is normally equipped with a 4- or 6-row boom, but special application booms or hand guns may also be used.

High-Pressure Sprayers

High-pressure sprayers are used primarily for treating vegetable and fruit crops. They provide the high pressure required to break up the pesticide liquid enough to assure complete coverage of the fruit and foliage.

High-pressure sprayers are similar in design to multipurpose sprayers but have piston pumps. The pumps can deliver up to 100 liters/min at 400–500 psi. All hoses, pressure relief valves, nozzles, and valves must be high-pressure types. Field booms are designed for spraying of specific crops, such as tomatoes, grapes, and blueberries. Hand guns can be used for spraying large fruit trees.

Airblast Sprayers

Sprayers that use air as a vehicle to transport spray droplets from the point of release to the point of application are called *airblast* sprayers (Fig. 7.9). High-volume fans supply the air; adjustable air vanes permit spraying on one or both sides of the sprayer as it moves between rows of trees or across a field. Nozzles operating at low, moderate, or high pressure deliver the spray droplets into the high-velocity air stream. The high-speed air aids in breaking up larger droplets and transporting these smaller droplets for thorough coverage. Agitation of the spray material in the tank is usually accomplished with a mechanical agitator.

Low-Volume Air Sprayers (Mist Blowers)

A special class of air sprayers, called low-volume sprayers or *mist blowers,* is characterized by the following features:

- High air velocities (120–200+ mph) and somewhat lower air volumes than conventional airblast sprayers.
- Metering devices, which may or may not be conventional nozzles, that operate at low pressures and depend on the high-speed air for liquid breakup.
- Use of much lower volumes of water than conventional airblast sprayers.

A considerable saving in time and labor is possible with low-volume sprayers because less water is handled. However, equipment calibration is more critical, favorable weather for spraying is more essential, and pesticide coverage on some crops may be less satisfactory with low-volume sprayers than with conventional airblast sprayers.

Ultra Low Volume (ULV) Sprayers

ULV spraying applies a chemical concentrate directly without the use of water or any other liquid carrier. Many ULV ground sprayers use a fan that delivers high-speed air to help break up and transport the spray droplets. The main advantage of ULV sprayers is the labor and time saved due to the elimination of water. The main disadvantages are increased risk from spraying concentrate and the limited number of pesticides that are cleared for ULV application.

Dusters

Although dusters were once very popular for applying pesticides, their general use has declined in favor of sprayers. It is difficult to get uniform coverage of foliage with dusts, and they drift more than sprays.

Hand dusters, like hand sprayers, are for professional use in structures and can be used for spot treatments, gardens, and other small jobs. They may consist of a squeeze bulb, bellows, tube, or shaker, a sliding tube, or a fan powered by a hand crank.

Power dusters use a powered fan or blower to propel the dust to the target. They range from knapsack or backpack types to those mounted on or pulled by tractors. Their capacity in area treated per hour compares favorably with some sprayers.

A power duster should be easy to clean and give a uniform application rate as the hopper is emptied. With both hand and power dusters, the dust cloud should fall well away from the user.

Both dusters and granular applicators are speed sensitive, so uniform speed must be maintained without traveling too fast for ground conditions. Bouncing equipment will cause the application rate to vary. It is important to stay out of any dust cloud that may form.

Banders should be watched to see that bandwidth stays the same. Small height changes due to changing soil conditions may cause rapid changes in bandwidth. Equipment should be cleaned as directed by the operator's manual.

Spray Equipment Parts

The overall quality of a sprayer and its suitability for various uses depend upon the quality and type of its component parts. The critical parts of a sprayer, and their characteristics, are discussed in this section. Figure 7.10 illustrates the parts of a typical boom-type sprayer.

Fig. 7.10. Diagram of boom-type sprayer showing major parts and accessories.
Courtesy Am. Assoc. Vocational Instructional Materials

Tanks

Tanks should be made of rust-resistant steel, fiberglass, or polyethylene to avoid rust, sediment, plugging, and restriction problems. Tanks from 220 to 600 liters are available on mounted sprayers; 800-liter or larger tanks come on pull-type sprayers. Only one tank per spray rig is recommended because more than one tank makes agitation difficult. Tanks need a large covered opening in the top with a removable strainer to make filling, inspection, and cleaning easy. A drain plug in the bottom is necessary for complete drainage during cleaning of the tank.

Tank size should be large enough to keep the refill frequency to a minimum. A calibrated sight gauge on the tank makes it easy for the operator to check the level of fluid at any time. A plastic tube is frequently used for this purpose. Plastic and fiberglass tanks may have calibrated markers on the outside of the tank. The fluid level is visible through the tank wall.

Pumps

The pump is the heart of the sprayer. It must have a pumping capacity of sufficient volume to supply the nozzles and hydraulic agitator, and to maintain the desired pressure at this full capacity. Pump parts must be able to withstand the corrosive effects of the spray materials and have a reasonable life.

The following six types of power-operated pumps are commonly used in pesticide sprayers:

1. *Piston* (reciprocating)—On upstroke, the piston pulls liquid into the pump chamber. On the opposite stroke, the fluid is forced out of the chamber through a surge tank to allow for a continuous and steady flow.
2. *Diaphragm*—Similar to a piston pump except the piston is replaced by a flexible diaphragm.
3. *Roller-Impeller*—Rollers are fitted into slots formed on the periphery of the impeller (a solid wheel fitted onto a shaft). The slots allow the rollers to follow the off-center contour of the pump housing. Fluid fed into the inlet side is forced out by the squeezing action of the rollers.
4. *Flexible-Impeller*—Similar to roller-impeller pump except it has a series of rubber paddles attached to an off-center shaft. As the paddles turn they force the liquid out of the pump.
5. *Gear*—Fluid enters pump and gear teeth force liquid into discharge outlet.
6. *Centrifugal*—Liquid enters through the center of the rotating impeller and is flung outward by centrifugal force and forced out the outlet.

When wettable powders are used, the resistance of the pump to wear is very important. A pump that is subjected to moderate wear such as a roller pump should be sized to provide a reserve capacity about 50% greater than needed for the nozzles and agitator. This reserve capacity will extend the life of the pump.

A pump will be damaged if it is operated dry, or with a restricted inlet. Dry operation occurs when the pump is operated with the tank empty, or if it is operated unprimed. Pump damage from not being primed is reduced when the suction line enters the tank at the bottom. A pump may also be damaged by improper mounting. If the pump is mounted directly on the tractor PTO shaft, do not attempt to bolt it down. A torque bar that extends about 15 cm from the pump is chained to a stationary part of the tractor to prevent the pump body from rotating. With piston pumps two torque bars should be used. One is chained stationary, and the other is attached to a spring on the opposite side of the pump from the chain to keep the chain tight.

Table 7.2 lists the capacities, pressure ranges, wear characteristics, and other information that can be used to select the proper sprayer pump.

Agitators

Agitation is essential for a good uniform spray supply rate. Agitation can be accomplished by *mechanical* or *hydraulic* means. For large orchard and vegetable sprayers a mechanical paddle mounted lengthwise in the bottom of the tank and driven by power is the most efficient agitator (Fig. 7.11).

Most field sprayers have hydraulic agitation of the *bypass* or *jet* type. The bypass type uses the return hose from the pressure relief valve to agitate the liquid. The bypass agitator hose must extend to the bottom of the sprayer tank. Bypass agitation is not sufficient on field sprayers with tanks larger than 220 liters unless a centrifugal pump is used.

Gear, piston, and roller pumps require a jet agitator on 440-liter or larger tanks. The jet must be located in the bottom of the tank and held in place on a pipe (*never on a hose*). The jet agitator must be connected to the pressure side of the sprayer and must operate at the operating pressure of the sprayer. Do not attach jets to the bypass line. A 220-liter tank requires one or more jet outlets, 400- to 600-liter tanks should have three or more outlets on a manifold. The manifold should be placed horizontally about 5–10 cm from the bottom of the tank. The tank capacity and operating pressure will determine the minimum jet size.

The orifices in a jet agitator must be sized so that the pump can maintain the operating pressure when all the nozzles are operating. In addition, some reserve capacity is needed to compensate for pump and nozzle wear.

Fig. 7.11. Paddle-type mechanical agitator.
Courtesy Am. Assoc. Vocational Instructional Materials

Hoses

Hoses should be of neoprene or other oil-resistant materials and strong enough to tolerate peak pressures. Hose test pressure should be twice the operating pressure. The suction hose should be made of two-ply fabric. It

Table 7.2. *Typical Sprayer Pump Characteristics*

Pump type	Available construction material	Available port sizes (in.)	Maximum practical pressure (psi)	Capacity of $\frac{3}{4}$-in. port[a] (gal/min, 40 psi)	Self-priming	Effect of abrasive material	Reparable
External gear	Bronze Cast Iron	$\frac{1}{2}$–1	75	6	Partly	Severe	No
Internal Gear	Cast iron	$\frac{3}{4}$	75	6	Partly	Moderate	No
6 roller-impeller	Cast Iron Ni-resist	$\frac{3}{4}$–$1\frac{1}{2}$	75	6–8	No	Moderate	Yes
7 or 8 roller-impeller	Ni-resist Cast iron Bronze	$\frac{3}{4}$	75	10	No	Moderate	Yes
Centrifugal	Cast iron	$1\frac{1}{4}$–2	50	40[b]	No	Little	Maybe
Diaphragm	Cast iron	$\frac{3}{4}$	75	5	Yes	None	Yes
Piston	Cast iron	$\frac{3}{4}$–$1\frac{1}{4}$	400	6–10	Yes	Little	Yes

[a]Pump at 540 revolutions per minute (RPM).
[b]$1\frac{1}{4}$-in. suction × 1-in. discharge centrifugal pump. Pump RPM = 3250 at 50 PTO.

should be larger than the pressure hoses (boom, jet agitator, and bypass) because it must provide the total pump flow through the suction hose, suction strainer, fittings, and valves. The pressure drop in pressure hoses, fittings, valves, and strainers should not exceed 10% of the total operating pressure. Hose clamps of the worm-screw type are the most reliable and are required when the operating pressure exceeds 100 psi. When the pressure exceeds 300 psi, use two worm-screw clamps or other special clamps.

Keep hoses from kinking or being rubbed. Rinse them often, inside and outside, to prolong life. Remove and store hoses during off-season, or at least store the sprayer out of the sun. Always replace hoses at the first sign of surface deterioration.

Strainers

Strainers (filters) in sprayers prevent foreign materials from entering and wearing the precision parts of the sprayer; they also prevent nozzle tips from clogging. There should be a 12- to 25-mesh screen strainer in the filler opening, a 15- to 40-mesh screen strainer in the suction or supply line to the pump, and a 50- to 100-mesh screen strainer between the pressure relief valve and the boom. A 25- to 100-mesh screen strainer is needed with some nozzle assemblies. Strainers should be 50 mesh or larger for wettable powder applications.

The strainer screen should be large enough to prevent excessive pressure drop and plugging. *The suction strainer should have at least 6 cm² of screen area per liter per minute of flow.* Clean strainers after each use. Replace them if deterioration is evident. Strainers are the best defense against nozzle and pump wear and nozzle clogging. Use nozzle screens as large as nozzle sizes permit. The screen opening should be smaller than the nozzle opening.

Pressure Gauges

Pressure gauges are important because pressure affects the amount of spray applied and droplet sizes. Use the lowest pressure possible for each job because high pressure uses extra power, wears equipment faster, and increases drift. Pressure gauges should be handled with care and equipped with a gauge protector when spraying corrosive materials or when using a piston-type pump. Select a gauge that can be read accurately at the pressure to be used.

Check frequently for accuracy against an accurate gauge. Do not use a gauge under too much pressure. Keep glass faces clean and intact.

Pressure Regulators

Pressure is regulated with a pressure relief valve that opens when the operating pressure exceeds its setting. For low-pressure spraying (10–40 psi), a diaphragm pressure relief valve is recommended.

When the desired operating pressure on a sprayer cannot be reached, check the pressure relief valve by observing the flow from the return hose. If there is flow in the return hose before the operating pressure is reached, the relief valve is not seating and must be checked for wear, trash, and broken springs. Repair or replace defective relief valves IMMEDIATELY.

On units with a centrifugal pump, the pressure may be regulated with an adjustable orifice valve (*globe* or *gate valve*). This device will make the sprayer a constant-rate sprayer (that is, spray rate is independent of ground speed) during reasonable changes in ground speed as long as the tractor gear is not changed.

Cutoff Valves

There should be a quick-acting cutoff valve (or valves) on every sprayer between the pressure regulator and the nozzle. One valve may be used to cut off the entire operation, or a combination of two or three valves may be used to control the flow to different sections of the boom. Flow cutoff valves should be easily reached from the operator's seat.

Nozzles

Selection of nozzles that meet the requirements of each particular spray application is important in the effective use of agricultural chemicals. Nozzles help control the rate, the uniformity, the thoroughness, and the safety of pesticide application.

Select nozzles that will provide the proper particle size and application rate within the recommended range of pressures. The spray pattern and nozzle location should provide excellent coverage of the target area but allow little spray to escape to other areas. *Be sure that all nozzles on a boom are the same and the recommended type for the job.*

To meet various spray requirements, nozzles are classified by the spray delivery patterns, spray angle, discharge rate, and the materials from which they are made. Eight types of nozzles, based on their spray patterns, are common:

1. *Flat fan nozzles* (Fig. 7.12A) are used for broadcast or boom spraying. Drift is less with them than it is with standard cone nozzles. Because their spray rate tapers at the edges, these nozzle patterns must overlap 30–50% for even distribution.

2. *Even flat fan nozzles* (Fig. 7.12C) are used for band spraying with a uniform spray pattern throughout. They should not be used for broadcast spraying.

3. *Whirlchamber (nonclog) nozzles* are available as wide-angle (120°) hollow cone nozzles. Clogging is minimized and less drift occurs due to the lower boom height and large droplet size. Coverage remains fairly constant with changes in boom height.

A

B

C

D

E

F

Fig. 7.12. Spray patterns of various nozzle types. A—Flat fan. B—Hollow cone. C—Even flat fan. D—Solid cone. E—Boomless. F—Flooding. *Courtesy USDA*

4. *Flooding nozzles* (Fig. 7.12F) may be used to apply herbicides and fertilizer solutions. They operate at low pressures and have wide spray patterns (up to 160°). The pattern width varies with pressure and height.

5. *Boomless (or cluster) nozzles* (Fig. 7.12E) are used for wide swath spraying. Only one or a cluster of nozzles may be used. The boomless nozzle does not give as uniform coverage as other types, and the spray pattern is more affected by wind than a boom-type nozzle.

6. *Hollow cone nozzles* (Fig. 7.12B) are designed for moderate to high pressures and are used to obtain thorough coverage of crop foliage and very uniform distribution.

7. *Disk–core cone nozzles* are used to produce hollow or solid cone (Fig. 7.12D) spray patterns. The combination of disk and core used determines the spray pattern and angle. These nozzles are used for spraying abrasive materials at high pressures.

8. *Solid cone nozzles* (Fig. 7.12D) are used for hand spraying, spot spraying, and foliar application of pesticides at moderate pressures.

The correct operating pressure should be selected for each spray job. Since drift is often a problem when spraying herbicides, the pressure should be as low as possible for the nozzle to operate. Insecticides are normally sprayed at moderate pressures; fungicides are sprayed at high pressures to achieve adequate coverage. Table 7.3 lists the type of nozzle and operating pressure recommended for various types of spraying.

Table 7.3. Nozzle Types and Operating Pressures for Different Uses

Weeds	Insects	Diseases
Flat fan, 20–40 psi	Flat fan, 20–60 psi	Flat fan, 40–60 psi
Wide angle flooding, 10–40 psi	Hollow cone, 40–120 psi	Hollow cone, 60–120 psi
Whirlchamber, 7–20 psi	Disk cone, 40–100 psi	Disk cone, 80–400 psi

Spray nozzles are made of the following materials for the reasons indicated:

- *Brass*—commonly used, inexpensive.
- *Stainless steel*—noncorrosive, expensive.
- *Aluminum*—resistant to moderately corrosive chemicals.
- *Tungsten carbide tips*—used for very abrasive materials such as wettable powders.
- *Plastic*—used for nonabrasive materials, corrosion resistant, inexpensive.

During spraying operations, nozzles should be watched for clogging and changes in nozzle patterns. If nozzles clog or other trouble occurs in the field, be careful not to contaminate yourself while correcting the problem. Shut off the sprayer and move it to the edge of the field before dismounting. Wear protective clothing while making repairs.

Pesticide Application Guidelines

Preapplication Checklist

All applicators should do the following *before* making any pesticide application:

- Identify the disease or pest.
- When in doubt, always ask the advice of an authority about pests and pesticides.
- Use only recommended pesticides for the problem.
- Use the least hazardous pesticides at the lowest rate that will do the job.
- Read the label.
- Know the signs and symptoms of pesticide poisoning.
- Know what to do in the event of an accident.
- Have their physician's phone number posted by their phone. In an emergency, time is extremely important.
- Take time to explain the safe use of all pesticides to employees. Make sure they understand.
- Users of organophosphorous and carbamate pesticides should contact their physician before using these materials and request a blood test. Inform the physician about the types of pesticides that will be used and make sure he has the current phone number for the nearest Poison Control Center.
- Check application equipment for proper operation.
- Calibrate equipment frequently for proper output (see Calibration of Equipment section).
- Check respirators for cleanliness and proper fit.
- Check gloves and other protective clothing before each use.
- Make sure plenty of clean water, soap, towels, and clean clothing are readily available in case of an accident.
- Have a responsible person receive pesticides and properly store them.
- Warn people in the area that pesticides are to be used and see that livestock and pets are removed from the area to be treated.
- Do not apply pesticides to food, feed, and water containers.
- Check to make sure that the time intervals between date of application and harvest will comply with those given on the label.

Ground Application

Ground applicators should do the following *before* and *during* any application:

- Make sure equipment has been properly calibrated (see Calibration of Equipment section).
- Mix only the amount of chemical necessary for the job.
- Use proper precautions for the chemical being used.
- Have two people working if highly toxic materials are being used.
- Stay out of drift or dripping foliage.
- Wear appropriate protective clothing.
- Inform workers to stay out of the treated area until the reentry time is up.

Fumigation

Fumigators should do the following *before, during,* and *after* any application:

- Make sure equipment has been calibrated (see Calibration of Equipment section).
- Post property before a fumigation.
- Have available and *use* a fumigation safety kit, an effective gas mask for the fumigant to be used, and a chart of instructions for artificial resuscitation.
- Take all safety precautions necessary to protect the public.
- Keep records of fumigation jobs.

Greenhouse Applications

Applying pesticides in greenhouses presents special problems. In a normal greenhouse operation, employees must work inside and personal contact with plants and other treated surfaces is almost a certainty. In addition, unauthorized persons may attempt to enter the premises. Ventilation in greenhouses is frequently kept to a minimum to maintain desired temperatures; as a result fumes, vapors, and dusts may remain in the air for considerable periods of time after pesticide applications. Pesticides applied to plants and other surfaces in greenhouses do not generally break down as rapidly as outside since they are not exposed to environmental factors such as rain and UV light.

The following precautions should be observed whenever pesticides are used in a greenhouse:

- Select the least hazardous chemicals that are effective for managing the particular pest problems encountered.
- Use gas masks and waterproof protective clothing as specified on the label when using hazardous pesticides.
- Post warning signs on the outside of all entrances when fumigants or other highly toxic pesticides are being applied.
- Read and follow label instructions carefully.
- Do not enter a treated building without a gas mask until it has been aired for the length of time recommended on the pesticide label.
- Avoid skin contact with treated plants and other surfaces for the time specified on the label. Where this is impossible, workers should wear protective clothing and wash frequently.

Postapplication Checklist

The following should be done *after* any pesticide application:

- Do not enter premises until the proper time interval has elapsed.
- Clean application equipment thoroughly (see Cleaning and Storing Equipment section).

- Clean and store protective equipment after each use.
- Change clothing and wash after using a pesticide.
- Store unused chemicals properly in original container with the label intact.
- Dispose of unwanted pesticides and containers properly.
- Keep accurate records on pesticide use.

Pesticide Use Records

Pesticide records can help users protect themselves and their investments. Records can help in the following ways:

- To determine what crop can be planted after certain herbicides and chlorinated hydrocarbon insecticides have been used.
- To improve pest control practices by comparing results obtained with various chemicals or equipment used.
- To plan for future pesticide needs and reduce carryovers.
- To establish proper use in case of residue questions in marketing your crop.
- To establish proof of use of recommended procedures if indemnity payments are involved.
- To determine if errors were made.

The following information should be kept when pesticides are used: crop and cultivar treated; pest or pests treated; location and acreage treated; time of day, date, and year of application; type of equipment used; pesticide used, amount used per hectare, stage of crop development; pest population (low, moderate, high); weather (temperature, wind, rainfall); harvest date; and results of application.

Keep notes in a pocket notebook. Write down information as it happens. Transfer it later to a permanent record sheet.

Cleaning and Storing Equipment

Sprayers

A sprayer should be thoroughly flushed and washed at the end of each day. *Never leave unused pesticides in the sprayer.*

Screens in the sediment chamber and nozzles should be removed and cleaned frequently with a small brush (a toothbrush will usually do). In some cases this may need to be done daily. Wire or nails should *never* be used to unplug nozzles because they may scratch the precision surfaces and distort the spray pattern. Use a wood splinter, if necessary, to remove an obstruction.

The pump, tank, and all fittings should be thoroughly washed several times with detergent and water after each spray job. Flush out the tank and then pump clean water through the pump and discharge equipment. Flush and drain the booms, hoses, and manifolds. Disassemble and clean nozzles. Nozzles may be stored in a can of light oil to prevent corrosion.

For long storage, a coating of oil should be "floated" onto the inside surfaces of the tank, pump, and piping as follows: Pour some new or used oil into the tank, and finish filling with water. Start the pump, and discharge the contents of the tank. A uniform coating of oil will remain on all interior surfaces to prevent corrosion. Drain all hoses; remove and fill the pump with permanent-type antifreeze to prevent freezing and sticking.

Store sprayers under a shed or in a building.

Granular and Dust Equipment

Dusters and granular equipment should be emptied and cleaned at the end of each day. Carefully estimate the material needed to avoid excess material in the equipment at the end of the job. Dusts and granules will cake or sometimes cause equipment to rust if left in equipment. After removing all material, clean the equipment according to the operator's manual; then oil or grease the bearings to prevent corrosion. Store the equipment in a dry place.

Fumigators

After using a fumigator, place the remaining fumigant in the original container and seal. Flush all pumps, orifices, hoses, and injectors and drain thoroughly. Kerosene or similar materials should be added as specified by manufacturer. Store the fumigant in a safe, cool, well-ventilated place. Store the equipment in a dry place.

Calibration of Equipment

Calibration—the process of adjusting equipment to apply a specified rate of material—is necessary to assure that each pesticide is applied as directed on its label. Too much pesticide is dangerous; too little will not do a good job. Only by calibrating correctly can efficient results be obtained. Careful attention to calibration procedures will prevent serious errors in application rates.

There are many ways to calibrate pesticide application equipment. The preferred methods differ according to the kind of equipment being used. Extension service personnel and firms that sell applicators can provide instructions about how to calibrate equipment. Additional methods for calibrating sprayers are also found in a guide prepared by the American Society of Agricultural Engineers.

Factors Affecting Application Rates

The rate of application of a pesticide depends on the application width, the discharge or flow rate of material from the applicator, and the ground speed. The application width, in turn, depends on nozzle spacing and row spacing. The flow rate is affected by nozzle size, as well as by the pressure on the nozzle and the density and viscosity of pesticide being applied, as follows:

1. The *pressure* on a nozzle or orifice influences its output rate but only in

proportion to the square root of the pressure. This means that the pressure must be increased four times to double the output (in other words, doubling the pressure, increases output by only 41%).

2. The *density* (weight per liter) of liquid affects the discharge rate through a nozzle or orifice at a given pressure. The heavier a liquid the slower it is discharged at the same pressure. Most nozzles are rated for water, which has a density of 1 kg/liter. When applying lighter (oils) or heavier (fertilizers and fumigants) liquids, adjustments must be made for the density of the material.

3. The *viscosity* or thickness of a material affects the flow rate. Thick liquids flow much slower than water. Most liquid pesticides have a viscosity very close to water, and viscosity is usually not a factor.

Ground speed is an important variable affecting the application rate of most equipment. For equipment that has a constant discharge rate (constant-pressure sprayers), the application rate is inversely proportional to the speed; that is, when the applicator is driven twice as fast, the application rate is reduced by one-half.

Application equipment can be calibrated to deliver a specified rate of material by adjusting one or more of the variables that affect application rate. Before calibrating any piece of equipment, set it up to do the desired job according to the manufacturer's instructions or other information sources.

Sprayer Calibration Methods

Before calibrating or using a sprayer, fill the tank with water and operate the sprayer to make sure that all parts are working properly. Remove all nozzles, boomcaps or plugs and thoroughly flush out the system. Then select nozzles of the proper type and size for the job. Check the discharge rate of each nozzle using a watch and measuring container. If the discharge rate of a particular nozzle varies more than 10–15% from the average, replace the tip.

There are two basic ways of calibrating a sprayer—spray methods and collection (pint jar) methods—although numerous variations are possible.

Spray a Hectare (Acre) Method

Probably the most accurate calibration method is to spray a measured hectare or acre and measure the quantity of material applied. This method can be used for all large acreage spray jobs. Calibration should be done in the field that will be sprayed, not on a road or other hard surface, because a sprayer will travel faster on a road than in a field at the same throttle setting. The following steps explain how to calibrate a sprayer by this method:

1. Completely fill the tank with water.
2. To determine how far to drive in the field, use one of the following formulas:

$$\text{Distance to drive for 1 ha} = \frac{10,000}{\text{sprayed width (m)}}$$

$$\text{Distance to drive for 1 acre} = \frac{43,560}{\text{sprayed width (ft)}}$$

Sprayed width is the distance between the nozzles on each end of the boom plus one between-nozzle spacing ($\frac{1}{2}$ of a nozzle spacing on each end of the boom).

3. Spray the measured hectare or acre at the desired speed and pressure.
4. Refill the tank carefully measuring the quantity required to fill the tank—this amount is the application rate per hectare or acre.
5. If the measured application rate is different from the desired rate, correct by changing speed and/or nozzle size.

Collection (Pint Jar) Methods

The pint jar methods are used to calibrate sprayers for row and broadcast ground application. In these methods of calibration 1 pt (473 ml or 16 oz) of liquid is collected from one nozzle. In Method 1 pesticide is collected from the sprayer while it is parked, whereas in Method 2 material is collected while the sprayer is moving.

Pint Jar Method 1

1. Select a nozzle with an average discharge rate.
2. Determine the average nozzle spacing (effective boom width divided by number of nozzles).
3. Measure the time it takes to collect 1 pt (473 ml or 16 oz) of water from the average nozzle while parked.
4. Beginning about 6 m (20 ft) behind a marker, drive in the field at the speed to be used when spraying. Begin timing the sprayer (tractor) when it passes the marker and drive for the same length of time that was required to collect 1 pint of water when the applicator was parked.
5. Measure the distance between the beginning and end markers. Compare this distance to those in Table 7.4 to determine the application rate.
6. If the rate is not acceptable, change nozzles and/or speed and recalibrate.

Pint Jar Method 2

1. Select a nozzle with an average discharge rate, and securely attach a pint jar under the nozzle to collect all the material from that nozzle. A jar lid with a hole for the nozzle will help retain all the material in the jar.
2. Starting at a marker, operate the sprayer at the desired speed and pressure until 1 pt (473 ml or 16 oz) is collected from the one nozzle.
3. Measure the distance required to fill the pint jar.
4. Determine the average nozzle spacing (effective boom width divided by number of nozzles).
5. Compare this distance with those in Table 7.4 to determine the rate.
6. If the rate is not acceptable, change the speed and/or nozzles and recalibrate.

Table 7.4. Calibration Table for Sprayers (Pint Jar Method)[a]

Nozzle spacing[b] (in.)	Distance required to catch 1 pt per nozzle at various application rates[c]							
	5 gal/A	7½ gal/A	10 gal/A	12½ gal/A	15 gal/A	20 gal/A	25 gal/A	35 gal/A
6	2178 ft	1452 ft	1089 ft	871 ft	726 ft	545 ft	436 ft	311 ft
8	1634	1089	817	653	545	408	327	233
10	1307	871	653	523	436	327	261	187
12	1089	726	545	436	363	272	218	156
14	933	622	467	373	311	234	187	133
16	817	545	408	327	272	204	163	117
18	726	484	363	290	242	182	145	104
20*	653	436	327	261	218*	163	131	93
21	622	415	311	249	207	156	124	89
22	594	396	297	238	198	149	119	85
24	545	363	272	218	182	136	109	78
30	436	290	278	174	145	109	87	62
36	363	242	182	145	121	91	73	52
42	311	207	156	124	104	79	62	44
48	272	182	136	109	91	68	54	39

[a]To use this table, find the value of the average nozzle spacing of the sprayer being calibrated in the left-hand column, then read across the row to the value closest to the distance traveled in the test procedure; finally read up the column to find the corresponding application rate in gallons per acre. For example, if a sprayer has a nozzle spacing of 20 in. and travels 218 ft during collection of 1 pt of material (or water) from one nozzle, the application rate is 15 gal/acre. The speed is accounted for in the distance.

[b]When nozzle spacing is not uniform or when more than one nozzle is used per row, use the average spacing. If three nozzles are used per row and the row spacing is 42 in., the nozzle spacing would be 42 ÷ 3, or 14 in.

[c]To convert rate to liters per hectare, multiply the table value by 9.3.

Fumigator Calibration

Fumigants are metered from an applicator with orifices. These orifices serve the same purpose as spray nozzles except that they are placed in the line going to the injector. The orifice size is selected in the same manner as a nozzle. First, determine the required flow rate in liters or gallons per hour, or liters or gallons per minute, using a spray nozzle formula. Then, using an orifice (spray nozzle) catalog or a flow formula, determine the size orifice needed.

Low-Pressure Fumigators

A low-volatile fumigator may be calibrated by collecting the material applied over a given area. There are two methods of collecting the fumigant: while the applicator is moving and while it is stationary.

The fumigant must remain as a liquid until it passes through the metering orifices. With low-volatile fumigators this is a problem only when the metering pump is operating at high speeds. Even though low-volatile fumigants are slow to vaporize, they will form vapor in the metering pump at high speeds due to the sudden vacuum created on the intake stroke. If this happens, the application rate is drastically reduced.

Method 1

1. Measure 30.5 m (100 ft) in the field.
2. With the applicator equipment running at the desired speed and engaging the soil, collect the fumigant applied in the 30.5 m (100 ft) section from one or more soil tubes (the entire applicator output may be collected).
3. Determine the application width being collected. For example, if the material is collected from 6 tubes spaced 20.3 cm (8 in.) apart, the application width is 121.9 cm (48 in.). If the material for one row is collected, the application width is the row spacing.
4. Compare the amount of material collected with those in Table 7.5 to determine the application rate.
5. If the rate is not acceptable, adjust orifices and/or pressure and recalibrate.

Method 2

1. Measure 30.5 m (100 ft) in the field.
2. While operating the equipment with the fumigant shut off, record the time required to travel 30.5 m (100 ft).
3. Determine the application width for the orifice or orifices to be used. For row application, use the row width and collect all material for one row.
4. While the application equipment is not moving, collect fumigant for the length of time required to travel 30.5 m (100 ft) in the field. Compare the quantity collected with those in Table 7.5 to determine the application rate.
5. If the rate is not acceptable, adjust orifices and/or pressure and recalibrate.

Table 7.5. Calibration Table for Low-Pressure Fumigators[a]

Application rate[b] (gal/A)	Quantity collected per 30.5 m (100 ft) at various application widths									
	61 cm (24 in.)		76.2 cm (30 in.)		91.4 cm (36 in.)		106.7 cm (42 in.)		121.9 cm (48 in.)	
1	0.6 oz	17.4 ml	0.7 oz	21.7 ml	0.9 oz	26.1 ml	1.0 oz	30.4 ml	1.2 oz	34.8 ml
3	1.7	52.1	2.2	65.0	2.6	78.2	3.1	91.2	3.5	104.3
5	2.9	86.9	3.7	108.4	4.4	130.3	5.1	152.1	5.9	173.8
7	4.1	121.6	5.1	151.7	6.2	182.5	7.2	212.9	8.2	235.1
9	5.3	156.4	6.1	195.1	7.9	234.6	9.3	273.7	10.6	312.8
12	7.1	208.5	8.8	260.1	10.6	312.8	12.3	365.0	14.1	417.1
15	8.8	260.7	11.0	325.2	13.2	391.0	15.4	456.2	17.6	521.3

[a]To use this table, find the column headed by the value corresponding to the application width of the fumigator being calibrated, then read down the column to the value closest to the amount of fumigant (in ounces or milliliters) collected during the test procedure; finally read *left* across the row to find the corresponding application rate.
[b]To convert rate to liters per hectare, multiply the table value by 9.3.

High-Pressure Fumigators

To calibrate high-volatile applicators dispensing methyl bromide, *the container must be weighed before and after applying fumigant to a measured plot.* The material cannot be collected because it would immediately volatilize. The following steps are involved in calibrating a high-pressure fumigator:

1. Weigh the container of fumigant.
2. Apply the fumigant to 30.5 m (100 ft) of row for row treatments or 93 m² (1000 ft²) of area for broadcast treatments.
3. Reweigh the container to determine the amount of fumigant used.
4. To calculate the application rate for row treatments, multiply the amount used (in pounds) per 30.5 m by the following values depending on row width:

Row width	Multiply by
30 in. (76.2 cm)	175
36 in. (91.4 cm)	145
42 in. (106.4 cm)	121
48 in. (121.9 cm)	105

5. To calculate the application rate for broadcast treatments in pounds per acre, multiply the amount used (in pounds) per 93 m² by 43.5; multiply by 107.5 to obtain the rate in pounds per hectare.
6. If the rate is not acceptable, adjust orifices and/or pressure and recalibrate.

Remember that speed, pressure, and applicator injector spacing will affect the rate of application.

The fumigant must remain as a liquid until it passes through the metering orifices. A highly volatile fumigant must be pressurized enough to prevent formation of vapor until after it passes through the metering orifices. The minimum pressure required to maintain a fumigant as a liquid depends on its temperature. The warmer the fumigant, the higher the pressure at which it will vaporize. If the fumigant forms a bubble, less material will pass through the orifice as a gas than as a liquid; this will reduce the rate.

Table 7.6 lists the minimum metering pressures required to maintain methyl bromide fumigants as liquids at different temperatures. This table is based on a pressure loss of about 2 psi in the system due to tube friction. If excessive

Table 7.6. Minimum Metering Pressure for Methyl Bromide Fumigators

Fumigant temperature		100% Methyl bromide	70% Methyl bromide 30% chloropicrin
(°C)	(°F)	(psi)	(psi)
10	50	5	0
20	68	15	5
30	86	25	15
40	104	40	25

pressure drop is present because of undersized fittings, tubes, etc., the pressures indicated in Table 7.6 must be increased.

When metering highly volatile fumigants at low pressure it is extremely important to keep the fumigant container cool (out of the sun) to prevent vapor forming in the metering system.

Granular Applicator Calibration

Granular applicators may be broadcast or row types. They are calibrated by collecting and weighing the material applied for a measured distance. A very accurate scale is required.

1. Collect the sample as follows:
   Row applicator—Collect the quantity applied per row for 30.5 m (100 ft).
   Broadcast tube applicator with spreaders—Collect the material applied per tube for 30.5 m (100 ft).
   Manual centrifugal (slinger) broadcast applicator—Place the spreader in a large paper or plastic bag and collect the material applied during operation for 30.5 m (100 ft) at normal speed.
   Power centrifugal (slinger) broadcast applicator—With the hopper empty, determine how long it takes to move 30.5 m (100 ft). Then fill the hopper with granules; with the applicator stationary, collect the material dispensed during the length of time required to move 30.5 m. If the applicator cannot be calibrated in a stationary position, collect the material in a suitable container while operating it over a specified distance.
2. Determine the application width (row spacing or tube spacing as appropriate).

Table 7.7. *Calibration Table for Granular Applicators*[a]

| Application rate[b] (lb/A) | Quantity collected per 30.5 m (100 ft) at various application widths | | | | |
	30.5 cm (12 in.)	61 cm (24 in.)	91.4 cm (36 in.)	106.7 cm (42 in.)	121.9 cm (48 in.)
5	5 g	10 g	16 g	18 g	21 g
10	10	21	31	36	42
15	15	31	47	55	62
20	21	42	62	73	83
25	26	52	78	91	104
30	31	63	93	109	125
40	42	83	125	146	167
50	52	104	156	182	208
60	63	125	187	218	250
75	78	156	234	273	312

[a]To use this table, find the column headed by the value corresponding to the application width of the applicator being calibrated, then read down the column to the value closest to the weight of the granules collected during the test procedure; finally read *left* across the row to find the corresponding application rate.
[b]To convert rate to kilograms per hectare, multiply the table value by 1.1.

3. Weigh the sample with a gram scale or very accurate ounce scale. (Do not include weight of container.)
4. Compare the weight of the sample collected with those in Table 7.7 to determine the application rate.
5. If the rate is different from the desired rate, adjust the equipment and recalibrate.

Protecting the Environment

Anyone who applies pesticides has an obligation for the safety of others and the environment. Our human environment is part of the whole *ecosystem*. This term describes the entire interacting system of animals and plants—of all kinds from bacteria to man—living in the physical world. Every plant or animal is affected by other plants and animals in the environment and by the physical characteristics of the environment.

Under a given set of conditions, the different species of plants and animals tend to come to an equilibrium among themselves, the so-called *balance of nature*. Pesticides are used to force the balance to be favorable to man, to provide him food and freedom from disease. Ecological theory says that the greater the number of species, and the greater the differences among these species, the greater the stability of the overall system and the less it will react to changes in the population of any one species.

Human activities often reduce the diversity of natural species. For example, in commercial agriculture a single crop is planted over a wide area, replacing the original mixture of many plant species. Such a system is not stable, and we must expect that it will never be easy to maintain a balance favorable to man. In the past we have found that some pesticides not only reduce the numbers of *target* pest species, as intended, but also change the numbers of such desirable *nontarget* forms as birds, with the result that the balance of nature was changed in unintended, and sometimes unfavorable ways. With man's need for food, we can never return to a truly natural condition. Instead we must learn how to reach the most reasonable balance we can, considering all human needs.

We cannot do much about controlling such things in the environment as light, rain, temperature, and wind, but we can control some other things, including the use of pesticides. Many people consider pesticides a tool for preserving or improving the environment. Others feel that pesticides cause pollution. As a weed is a "plant out of place," a pesticide sometimes can be a "tool out of place." Correct use prevents pollution by pesticides; incorrect use can lead to lawsuits, fines, and payment of damages by the applicator.

Controversies and problems related to the use of pesticides are likely to be greater when agricultural land lies close to urban areas than they are in all-agricultural areas. Pesticide damage to nontarget organisms can cause problems in any community, but the denser the human population in the area adjoining treated areas, the greater the risk of injury to humans and domestic pets. Furthermore, contamination of a reservoir used for a community's water supply and public recreation, by careless pesticide application, can cause major disruptions in a community and, quite possibly, severe sanctions against the offending applicator.

Pesticide Damage to the Environment

Using pesticides in a way other than as directed on the label can (1) injure plants and animals, (2) leave illegal residues, and (3) damage the environment. Any pesticide can cause harm if not chosen and used with care.

Not all pesticides act the same after application. Most are classified either as nonpersistent or persistent. *Nonpersistent* pesticides break down quickly and remain on the target or in the environment only a short time before being changed into harmless products. Some are highly toxic; others are fairly nontoxic. *Persistent* pesticides break down slowly and may stay in the environment without change for a long time. Often this characteristic is good for pest management because it provides long-term management with a single application. Most persistent pesticides are not broken down easily by microorganisms and are only slightly soluble in water.

Some persistent pesticides that can injure sensitive crops planted on the same soil the next year seem to be of little hazard to the environment beyond the treated soil. Other persistent pesticides can accumulate in the bodies of animals, including man. Such pesticides may build up to a level that is harmful to the animal itself or to a meat eater that feeds on it. These are called *accumulative pesticides.*

The chances that pesticides will cause environmental damage increase when they move off the target. This can occur by various means:

- Drifting out of the target area as a mist or dust
- Evaporating and moving with air currents
- Moving on soil through runoff or erosion
- Leaching through the soil
- Being carried out as residues in crops and livestock.

When the plants on land adjoining a treated area are sensitive to the pesticide being applied, it is very important that drift be controlled (see Controlling Pesticide Drift section).

Direct Kill of Nontarget Organisms

Pesticides are sometimes applied over a large area to such target organisms as fungi, insects, and weeds. Many nontarget plants and animals within the treated area may be susceptible to a pesticide and can be harmed if they come in contact with it. For example, herbicides may kill nearby crops and landscape plants; some insecticides are likely to kill any bees or other pollinators that are in a field during application. Runoff from a sprayed field can kill fish in a nearby stream or pond. Life in streams can be wiped out by careless tank-filling or draining and improper container disposal.

To reduce the danger to nontarget organisms, the most selective and least toxic pesticide that is effective for a given pest problem should always be chosen.

Contamination of Soil and Water

A number of factors affect the risk of soil and water contamination by pesticides. On crops where repeated pesticide application is necessary, pesticides (especially those that are persistent) may build up in the soil. On row crops where tilling is common and the water drains into aquatic areas, runoff of contaminated soil particles and pesticides is likely. Pesticides should not be applied just before predicted heavy rains or just before some types of irrigation.

Although some fungicides and herbicides are more effective when applied before a rain, heavy rains cause runoff and tend to wash the pesticide away from the target area. The runoff can carry the pesticide into sensitive areas where crop injury may result. Runoff may also reach farm ponds, streams, and waterways, causing contamination, fish kills, or injury to domestic animals such as dairy cows.

The use of persistent pesticides may limit future planting options. *Carryover* of certain pesticides in the soil from one year to the next means that only crops that the pesticide will not kill or contaminate can be planted. Even pesticides directed at plants or animals can move to the soil. Pesticides may be washed or brushed off. They may be worked into the soil with dead plant parts.

Many combinations of these and other factors may lead to problems associated with soil and water contamination by pesticides. As the risks of contamination increase, applicators may need to consider changing time of application, pesticides, or normal application method.

Controlling Pesticide Drift

Pesticide drift is the movement of spray or dust particles or chemical vapors a considerable distance from the place of release or discharge. Most frequently it is associated with agriculture and is considered undesirable because it can injure or leave undesirable residues on nearby crops and livestock, pollute the environment, and injure man. It also wastes chemicals and causes uneven distribution.

Spray or Dust Drift

A number of factors influence the movement of particles discharged from application equipment. The most important are *weather* and *particle size*. As a general rule, the application of pesticides should be avoided when the wind is blowing faster than 12.8 km/hr (8 mph). When spraying near the property of others or in public areas, be conscious of the possibility of drift and take necessary precautions. Do not spray near these areas on windy days.

A weather condition called *inversion*, where warm air overhead traps cool air beneath, can cause air movement and drift. Inversions usually occur in early morning or late afternoon.

Droplet size influences length of drifts. For example, with a 4.8 km/hr (3 mph) wind, visible spray particles (50 μm or more) may drift as much as 100 m (109 yd), whereas invisible vapor (10 μm or less) may drift several kilometers (a mile or more). Because of their light weight and small particle size, dusts often present a greater drift hazard than sprays.

Other variables that influence droplet size are *pressure* and *nozzle orifice.* Too often farmers assume that high pressures (100 psi, for example) will give more effective pest management by driving the spray material down into the leaves. However, high pressure reduces the size of spray particles causing more drifting; the higher velocity also causes more turbulence with less effective coverage. Many fungicides and insecticides must be applied at high pressure (100–400 psi) to reach and manage the intended pests. However, even where high pressures are needed, excessive pressure should be avoided to minimize drift. Increasing the nozzle orifice (opening) size can help reduce drift because small openings require more pressure, resulting in smaller droplet sizes and hence more drift.

A wide range of chemical formulations is available for spray applications, and the spray particle size can be altered to fit application needs. With the increasing popularity of low-volume and ULV spraying, drift problems will increase in number and intensity. However, the recently developed *foam* technique that applies pesticides in a foam is helpful in solving drift problems. Reported benefits of foam include extended application periods (wind no factor), less material used (more complete coverage, no overlapping), cleaner equipment, and fewer reloading trips (less water used).

Vapor Drift

Vapor drift, unlike spray or dust drift, is related directly to the chemical properties of the pesticide. Most problems are the result of accidental or unintentional misuse. If the applicator understands the chemical properties, he can control and avoid damage from drift.

Vapor drift can be due to vapor leakage. Vapor leakage can be stopped by properly sealing fumigant or other volatile materials after they are applied and by applying these materials with vapor-tight equipment. Soil sterilization in a greenhouse that ends up damaging plants in an adjacent greenhouse is one example of vapor drift. Ester formulations of phenoxy herbicides may volatilize and drift under high-temperature conditions. Amine or acid formulations should be used where vapor drift might cause problems.

Facts to Keep in Mind

When other conditions are the same, the following generalizations apply:

- Smaller droplets or dust particles drift farther than larger ones.
- There is more chance of drift with airblast sprayers than with boom sprayers.
- There is more chance of drift with high-pressure sprayers than with low-pressure sprayers.
- Low-volume concentrates are more likely to drift than high-volume dilutes.

8

Biotechnology

Through the centuries humans came to realize that certain animals could not crossbreed and produce living offspring. Farmers observed that horses could not breed successfully with cows, nor could swine breed with sheep. Likewise, they saw that only closely related plants could crossbreed and produce living seed: grapes could not be bred successfully with grasses, nor corn with beans.

However, farmers have known for hundreds of years how to select, breed, and improve domesticated animals and plants. The Arabian horse has been selected and bred for speed and endurance, the Clydesdale for pulling heavy loads. Beef cattle have been developed for meat production, dairy cattle for milk production. Similarly, farmers have cross-pollinated and propagated grapes to improve a certain desirable flavor and wheat with higher yields, improved quality and ability to withstand cold weather.

Throughout recorded history the domestication, breeding, and improvement of animals and plants has depended on the fundamental observation that living organisms both resemble and differ from their parents. Within the past century scientists have gained an increasing understanding of heredity and change among living things—the science of genetics, which has developed and expanded rapidly. It is now established that each inherited trait or function of an organism depends upon a unit of inheritance, a specific *gene,* which directs its development. In microorganisms (bacteria) that commonly reproduce asexually and whose single cells lack a defined nucleus, a single gene controls each specific trait. In higher plants and animals two genes exist with respect to each trait: one inherited from each parent.

Some plants have as many as 10,000 genes or more. Each gene is a tiny segment of a threadlike component of the cell called a *chromosome.* The chromosomes, located in the nucleus, are composed of deoxyribonucleic acid (*DNA*) and protein. DNA is a wondrous chemical that can reproduce exact copies of itself. Many genes make up one chromosome. Each gene is a tiny piece of DNA that regulates the reproduction of a specific protein, which in turn governs the composition and development of a specific trait. Thus, the fidelity and diversity of inherited traits are locked within the DNA of chromosomes.

By using the accumulated knowledge of the laws of genetics and the techniques of controlled crossbreeding and selection, plant breeders have greatly improved the usefulness of many farm crops. Breeders have increased crop yields and quality, incorporated more effective resistance to disease and insects into cultivars, expanded the geographical limits of crop production by increasing crop tolerance to stress conditions (cold, day length, drought, heat, salinity), and overcome natural breeding barriers such as sterility and abortion. The increases in crop yields in the United States between 1930 and 1975 are shown in Table 8.1. The degree to which applied genetics has contributed to these yield increases is impossible to determine because there have been great improvements in farm management, fertilizers, herbicides, irrigation, and pest management. In the case of corn, however, it is estimated that the use of hybrid corn accounts for as much as 50% of the harvest increases. Development of disease-resistant cultivars also accounts for a portion of the yield increases. By cross-pollinating a low-quality, scab-resistant apple cultivar with a high-quality, scab-susceptible cultivar and selecting through several generations, pomologists have developed a scab-resistant apple of high quality.

Table 8.1. *Average Yields of Major Crops in the United States—1930 and 1975*

	Yield per acre[a]			Increase (%)
	1930	1975	Unit	
Wheat	14.2	30.6	Bushels	115
Rye	12.4	22.0	Bushels	77
Rice	46.5	101.0	Bushels	117
Corn	20.5	86.2	Bushels	320
Oats	32.0	48.1	Bushels	50
Barley	23.8	44.0	Bushels	85
Grain sorghum	10.7	49.0	Bushels	358
Cotton	157.1	453.0	Pounds	188
Sugar beets	11.9	19.3	Tons	62
Sugarcane	15.5	37.4	Tons	141
Tobacco	775.9	2,011.0	Pounds	159
Peanuts	649.9	2,565.0	Pounds	295
Soybeans	13.4	28.4	Bushels	112
Snap beans	27.9	37.0	Hundredweight	33
Potatoes	61.0	251.0	Hundredweight	129
Onions	159.0	306.0	Hundredweight	92
Tomatoes				
Fresh market	61.0	166.0	Hundredweight	172
Processing	4.3	22.1	Tons	413
Hops	1,202.0	1,742.0	Pounds	45

Source: USDA (1979).
[a]To determine yields per hectare, multiply table values by 2.47.

Breeding a disease-resistant cultivar is usually a long and laborious process. After a source of resistance is found, 10 years or more may elapse before crosses are made and suitable cultivars selected. The incorporation of resistance to several diseases into a desirable agronomic cultivar becomes increasingly difficult and tedious. (See Chapter 6, Disease-Resistant Cultivars section.)

Each major agricultural crop has numerous relatives, both wild and cultivated, that contain many desirable characteristics. For example, corn is not susceptible to attack by the fungus that causes stem rust of wheat; all presently cultivated corn cultivars are immune to wheat rust. Obviously this immunity is inherited and transmitted from one generation to the next. If some way could be found to transfer rust immunity from corn to wheat and incorporate the resistance into usable cultivars, then wheat crops would be protected from attack by this pathogen, which annually reduces worldwide yields.

However, until the last few decades, transfer of desirable traits from one species of plant to another was possible only if they could be mated by cross-pollination. It is still impossible to crossbreed or hybridize unrelated plants such as soybeans and tomatoes by cross-pollinating them. In addition, many plant species are difficult to manipulate genetically by conventional methods. Therefore, more rapid and convenient ways to improve plants are being sought. Classical or conventional genetics will continue to play the major role in animal and plant breeding, but new genetic techniques will assume ever increasing importance.

Since 1960, scientists have made great progress in finding ways to genetically change and improve plants, including microorganisms (bacteria and fungi). This new technology, a form of *biotechnology,* deals with ways to change and/or grow plants in test tubes in the laboratory, rather than by use of conventional plant breeding techniques that involve crossbreeding. As a result, the development of superior strains of food crops and trees can proceed at a faster pace. Scientists hope to accomplish within a few years in the laboratory what usually takes 10–40 years to do by traditional methods. Biotechnology includes *genetic engineering,* by which the hereditary apparatus (DNA) of a microorganism is altered so that the cell can produce more or different chemicals or perform completely new functions. These altered cells are then used in industrial production.

Vegetative Propagation

For centuries many agricultural plants, including banana, Irish potato, strawberry, sugarcane, and others that usually do not produce true seed, have been routinely propagated vegetatively by rooting of cuttings, i.e., by *cloning.* Also, some economically important plants (bulbs, chrysanthemum, fruit trees, Irish potato, sugarcane) that are *heterozygous* and do not breed true must be propagated by vegetative means so that the desirable characteristics of each cultivar will be perpetuated. Cloning can be accomplished with (1) organs (flowers, leaves, roots, stems), (2) tissues (endosperm, epidermis, mesophyll, nucellus), and (3) single cells.

Organ Culture

When pieces of living organs (leaves, roots and stems) of some plants in the form of adventitious buds, cuttings, rhizomes, shoots, stolons, and underground stems are cut or broken off and placed in water or another suitable rooting material with favorable growing conditions, they often produce roots and eventually a new plant. This ancient process, sometimes called *organ culture,* has been widely used for hundreds of years. It is, however, laborious and time-consuming to grow plants from cuttings or rhizomes, even though conventional methods have improved since the 1940s by the use of rooting hormones, controlled light and mist.

Adventitious Shoots and Callus Tissue

Adventitious shoots are often formed on roots, lower parts of stems, or trailing leaves of some plant species (Fig. 8.1). If small bits of these plant tissues, called *explants,* are first dipped in alcohol or sodium hypochlorite to kill any contaminating bacteria or fungi, and then grown in flasks, they may produce many adventitious shoots that will grow into new plants. In other cases, the explant may form a cluster of cells, called a *callus,* that consists of an unorganized mass of dividing and enlarging cells. If culture conditions are readjusted when the callus appears, the cells can become specialized into tissues and organs and eventually into a complete normal plant. Each callus can be further subcultured. Routinely, as many as a thousand identical plants can be produced from each gram of original explant.

Immature embryos are well suited as donor tissues in the regeneration of cereal plants (corn, millet, oats, sorghum, and wheat). Embryos of these crops can be easily dissected out of the parent plant, are easily preserved in storage, and produce vigorous cultures that yield large numbers of regenerated plants. Thousands of "daughter" plants can be easily produced from a single embryo. Another effective and widely used donor tissue for grasses and cereals is the immature flower.

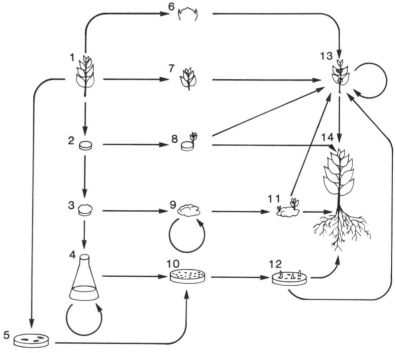

Fig. 8.1. Diagram showing relations between the various methods of *in vitro* propagation. 1—parent plant; 2—organ explant; 3—callus induction; 4—cell suspension culture; 5— liberation of protoplasts; 6—meristem culture; 7—shoot–tip culture; 8—adventitious shoot; 9—callus multiplication; 10—plated cells or protoplasts; 11,12—shoots from callus or cells; 13—axillary shoot proliferation; 14—rooted plantlets for field use.

Tissue and Single-Cell Culture

Many plants can be propagated from meristems, tissues, or single cells. However, both tissue and single-cell culture require more precise techniques than organ culture. Since 1960, increased scientific understanding and more refined laboratory techniques have led many commercial enterprises to concentrate on cloning crop and ornamental plants under controlled conditions. More than 300 species of plants are now propagated this way. A few examples are listed in Table 8.2.

Table 8.2. Some Plants Propagated through Tissue Culture for Production or Breeding

Floriculture and horticulture

Vegetable crops
- Asparagus
- Beet
- Brussels sprouts
- Cauliflower
- Eggplant
- Onion
- Spinach
- Sweet potato
- Tomato

Fruit and nut trees
- Almond
- Apple
- Banana
- Coffee
- Date
- Grapefruit
- Lemon
- Olive
- Orange
- Peach

Fruit and berries
- Blackberry
- Grape
- Pineapple
- Strawberry

Foliage
- Silver vase
- Begonia
- Cryptanthus
- Dieffenbachia
- Dracaena
- Fiddleleaf
- Poinsettia
- Weeping fig
- Rubber plant

Flowers
- African violet
- Anthurium
- Chrysanthemum
- Gerbera daisy
- Gloxinia
- Petunia
- Rose
- Orchid

Ferns
- Australian tree fern
- Boston fern
- Maidenhair fern
- Rabbitsfoot fern
- Staghorn fern
- Sword fern

Bulbs
- Lily
- Daylily
- Easter lily
- Hyacinth

Pharmaceutical
- Atropa
- Ginseng
- Pyrethum

Silviculture
- Douglas fir
- Pine
- Quaking aspen
- Redwood
- Rubber tree

Source: Office of Technology Assessment.

Meristem Culture

The shoot apex (growing point or *meristem*) of plants is the region that produces all the aboveground parts of the plant. If the meristem is cut off aseptically and placed in a nutrient broth or gel under proper conditions of light, temperature, and moisture, it will begin to grow. In some species, each meristem in culture grows into one plantlet. In others, the meristem forms multiple plantlets.

In small flasks or jars of about 250-ml ($\frac{1}{2}$-pt) capacity, meristems can be grown in large numbers on shelves in environmental chambers where light and temperature can be controlled. After the plantlets have grown large enough, they can be potted in ordinary potting soil and allowed to grow until they are large enough for use. The complete cycle from plant to meristem to plant permits plant production on a far more massive scale and in a far shorter time than is possible by conventional breeding techniques.

Cloning is particularly suited for securing daughter plants from valuable trees from which it is difficult to secure seed. Through meristem cultures and other cloning techniques, thousands of plants can be secured from a single "mother" tree.

Oftentimes valuable plant stocks used for propagation become infected with viruses that make them stunted, weak, or otherwise unsuited for commercial use. In former years these plants could not be used as parent material because new plants derived from such "mother" plants would also be infected with virus. It is now known that cells of the growing tips (meristems) of many virus-infected plants are free of virus. Culture of meristems from some virus-infected plants will result in new plantlets that are free of virus. In turn, meristems from these virus-free plantlets can be grown.

By using this method, growers have produced virus-free asparagus, banana, Irish potato, strawberry, sugarcane, and sweet potato. Several other crops including chrysanthemum, dahlia, lily, tobacco, and tulip have also been freed of virus by employing meristem culture. These virus-free meristems can be stored until needed in test-tube cultures or at freezing temperatures to protect them from reinfection. Another advantage of this method of propagation is that cultures of virus-free tissue can be sent to other countries and used immediately without having to be held in quarantine for long periods to establish that they are disease-free. Tissue culture provides an ideal method of storing and rapidly multiplying virus-free material.

Tissue Culture and Germplasm Preservation

The continuing search for ways to improve current cultivars and commercial crops makes preservation and maintenance of large collections of germplasm necessary. In this way, all characteristics of possible value can be saved for future investigation and study. Seed storage is a widely used and efficient storage method, but is not always feasible for the following reasons:

- Some plants do not produce seeds.
- Some seeds remain alive for a few months only.
- Some seeds are genetically mixed and not suitable for maintaining true-to-type breeding stocks.

- Seeds of certain species infected with seedborne pathogens die after a few months in storage.

Meristem and tissue cultures provide the scientist with additional ways to preserve germplasm. Cells, tissues, or small plantlets can be maintained at low temperatures or on starvation media. However, these techniques are not suitable for long-term storage of several years or more. As a result, storage at freezing temperatures has been investigated. Buds, embryos, meristems, plant cells, pollens, protoplasts, shoot tips, and whole seeds have been shown to remain alive when held at $-20°C$ in an ordinary home freezer, although progressive changes in quality occur. This deterioration can be prevented if a suitable protectant, such as dimethyl sulfoxide or glycerol, is added and storage is at a very low temperature, preferably that of liquid nitrogen ($-190°C$), which stops all living processes. This storage method is receiving increasing attention as a useful tool in germplasm preservation.

Regeneration of Plants from Cells

Using Protoplasts from One Species

It has long been known that normal plants of some species can be regenerated from single cultured cells. Animal cells have no rigid wall and can be grown directly without prior modification. Plant cells, on the other hand, contain rigid cellulose walls cemented together by *pectin* and hemicellulose material. Before individual plant cells can be made to grow and divide, the cell wall must be removed, usually by enzyme action. These naked cells, called *protoplasts,* may be isolated from flowers, leaves, roots, or stems from elite individual plants. Leaves are often used because they are the easiest tissues to convert into mass protoplast populations. By growth of protoplasts in aseptic baths of hormones, minerals, sugars, and vitamins, individual cells can be stimulated into growing into whole plants in the laboratory under controlled conditions.

Protoplasts have several characteristics that make them suitable for study:

- Large, relatively homogeneous populations of protoplasts, of known genetic composition, can be easily obtained from leaf mesophyll cells or other tissues from single plants.
- Protoplasts can be treated experimentally and cultured in the laboratory with high efficiency.
- The absence of a cell wall permits protoplasts to absorb genetic materials from the solution in which they are suspended.
- Protoplasts in solution can fuse with each other.

Researchers have noted that some tissue cultures derived from single protoplasts taken from virus-infected leaves are free of virus. This finding may indicate that, in addition to meristem cells, some viruses are unable to invade every cell in an infected leaf. Thus, protoplast culture may be used as a method to obtain virus-free tissue from desirable plants.

Protoplasts from several cereals and grasses have been induced to produce mature plants. Theoretically, all the protoplasts from the leaf of a plant should be genetically identical, but sometimes changes occur, or can be induced to occur. Tissues or plants derived from different protoplasts vary in their ability to resist or grow in the presence of toxins or herbicides or to absorb certain nutrients. These mutant plants with desirable or unusual traits can be identified and used.

Selected cells resistant to the toxin produced by race T of *Helminthosporium maydis* (Nisik. & Miyake), the fungus that causes southern corn leaf blight (SCLB), have been regenerated into mature plants. These plants are resistant to SCLB and can be used to produce seed resistant to the disease. In addition, plants of Irish potato grown from individual protoplasts proved to be resistant to late blight; others matured earlier and others produced more uniform potatoes. These results indicate that protoplast culture enables the scientist to direct plant development selectively to desired ends, and saves space and time.

Similarly, protoplasts from other plant tissues can be used for regeneration. Cells from fertilized eggs, embryos, endosperm, epithelial tissues or root cells can be manipulated and subjected to various stress conditions. As a result, desirable plants that require less water to grow or can withstand temperature extremes or high salinity can be selected.

Regeneration of Haploid Cells

The techniques described above all pertain to *diploid* cells—those with two sets of chromosomes: one set derived from the male parent and the other from the female parent. *Haploid* sex cells, those with only one set of chromosomes, can also be used for regenerating new plants. For example, cells from immature anthers or pollen cells containing one set of chromosomes can be isolated and regenerated. Also, cells from unfertilized eggs (ovules) can be grown in culture to produce haploid callus tissue and, eventually, haploid plants. This task is often a difficult one, but the addition of growth hormones (auxins, cytokinins, and gibberellins) at the correct time in correct amounts has made production of haploid plants possible. These plants then can be treated at the appropriate time with *colchicine* (a chemical that induces the cells to double their number of chromosomes) to produce diploid plants that are homozygous. In this genetic state, the plants can be used as new cultivars or as parental material to crossbreed with others to produce new cultivars with desirable characteristics. A new cultivar of barley was produced in 5 years by this method of doubling the chromosome number of haploid plants produced in tissue culture, and diploid tobacco plants derived from haploid plants obtained from anther tissue are routinely used as breeding material.

Fusing Protoplasts from Different Species

Often it is difficult to crossbreed distantly related plant species because natural barriers and sterility problems prevent sexual fertilization, embryo development, and/or seed formation. The fusion of asexual or vegetative cells to form hybrids and the subsequent regeneration of these hybrids into plants is an alternative method for the production of interspecific crosses in plants. Much information is available on cell fusion and/or somatic hybridization of plant cells.

Essentially, five steps are involved in this process:

1. Isolation of living protoplasts capable of division from the parent plant tissues.
2. Fusion of the two parental protoplasts.
3. Fusion and integration of the two parental nuclei with subsequent mitosis and formation of hybrid cells.
4. Detection and isolation of the hybrid cells.
5. Regeneration of tissues and then plants from hybrid cells.

Thus, it is possible to take a solution of protoplasts of, let us say, soybeans and mix them with protoplasts of sunflower. The mixture of protoplasts is then incubated in a nutrient broth under suitable conditions of light, temperature, and air, and in a few days time, some (a very low percentage) of the protoplasts will stick together or fuse. Even the nuclei fuse. The chromosomes from each "parent" pair, and subsequently normal mitotic divisions occur. The cell number increases with the chromosomes from both the soybean and sunflower contributing to the characteristics of the regenerated tissue which begins to form. After being allowed to undergo a few divisions, the cell mixture is diluted. Microdroplets are placed in small glass dishes with numerous small wells. Those wells containing single hybrid clusters can be scanned with a microscope, identified, and allowed to grow. By fusing protoplasts of tomato and potato, scientists have created a hybrid called a "pomato."

Unlike sexual crosses, the cytoplasms of the parental types become integrated during protoplast fusion. Because of this unique feature of somatic hybridization, certain heritable factors associated with the cytoplasm can be transferred. Transfer of male sterility, a desirable feature of economic importance in plant breeding, from one species of tobacco to another by protoplast fusion has been accomplished.

Successful protoplast fusion has been performed using barley and soybean, corn and soybean, potato and tomato, tobacco and soybean, and sorghum and corn protoplast mixtures, and the list of successes keeps growing. It is hoped that scientists will soon find ways to regenerate plants from these protoplast "hybrids." Somatic hybridization by fusion of plant protoplasts promises to be of great use in transferring genes governing resistance to viral diseases from one parent to otherwise susceptible species; indeed this has already been demonstrated. Somatic hybrid plants produced from the fusion of protoplasts of susceptible *Nicotiana tabacum* (tobacco) and resistant *N. rustica* (a relative of tobacco) were resistant to tobacco mosaic virus.

Transplanting Cell Organelles into Protoplasts

Although inducing entire protoplasts from different plant species to fuse is possible, it is still a rare event and difficult to accomplish. Isolation of a part of the protoplast of one cell and its transfer to another is much easier, and the physiological shock to the protoplast receiving the material is greatly reduced. Furthermore, even when protoplast fusion occurs, hybrid production may be precluded because of chromosome imbalance, loss of unpaired chromosomes, or lack of nuclear fusion.

To avoid some of the difficulties involved with cell fusion, scientists are conducting experiments with nuclei isolated from one kind of plant and injected or transplanted into the protoplasts of another kind. Here the nuclei begin to function and induce changes in the cells which reproduce to form tissues, and subsequently mature plants. The cytoplasm also contains other organelles, such as chloroplasts and mitochondria, floating as tiny blobs in the cytoplasmic "soup." It is well known that some genetic material is located in the cytoplasm. The inheritance of the following important traits has been linked with cytoplasmic genes:

- Male sterility
- Flower morphology
- Resistance to fungal toxins
- Organelle morphology
- Albinism and variegation
- Fraction 1 protein production
- Nitrate growth response

Many more cytoplasmic traits are yet to be determined.

Scientists have already successfully transplanted chloroplasts and mitochondria from one plant species into another. There are also reports that mitochondria from one kind of fungus have recombined with mitochondria from another fungus and produced a new type of "hybrid" fungus. Experiments like these offer hope that new and useful types of plants can be produced by transplanting organelles.

Transplanting DNA from One Cell to Another

An exciting bioengineering technique with important implications for agriculture involves the use of *recombinant DNA*—the hybrid DNA produced by joining pieces of DNA from different sources (parents). Scientists are improving this technique with the long-range goals of building plants that will be immune or resistant to disease, can extract nitrogen from the air, use sunlight more efficiently in photosynthesis, and have "foreign" genes that will improve crop yields and produce more nutritious proteins. Already scientists are constructing specific bacterial mutations at will and routinely producing millions of mutant cells for specialized uses and study. For example, it is now possible to identify and purify some genes, tailor them for insertion into a new bacterial cell, identify the cell, and clone the new cells in quantity to obtain a supply of DNA for future study and use.

Recombinant DNA technology (*gene splicing*) is sophisticated and elegant. It is based on the knowledge that an individual chromosome is a double helix (spiral) composed of thousands of *nucleotides* (the fundamental chemical units or building blocks of DNA) attached in a definite sequence (Fig. 8.2). Scientists are using two methods to splice genes into chromosomes. In the gene "tweezers" method, a solution of chromosomes from a desirable host is treated with a specific *restriction enzyme,* which has the ability to cut double-stranded DNA at a particular point (nucleotide sequence) where the desired DNA is attached. The snipped-out bit of DNA with chemically "sticky" ends is then treated with a *recombinant enzyme,* which helps stick the small piece of DNA or gene to a new chromosome. In other words, a gene from one chromosome is extracted and then inserted into the DNA of another species, i.e., a mutation has been engineered. In the "shot-gun" approach, several restriction enzymes are used to sever a large genome of one species into usable segments that can be easily spliced into the chromosome of another species.

Where would the selected genes for gene splicing come from? They might come from a closely related plant, another plant species, or from a bacterium—perhaps even from an animal. Theoretically, any gene that can replicate and function after being spliced into a recipient chromosome can be

Fig. 8.2. Deoxyribonucleic acid (DNA) consists of two intertwined strands resembling a twisted ladder. The four kinds of bases linking the strands always pair off in the same combinations: adenine–thymine (A–T) and cytosine–guanine (C–G). When a strand unzips, two new double strands exactly like the original are constructed.

used, for example, to confer disease resistance. Animal genes might be put into plants to produce crops with meatlike proteins.

Scientists also hope to use recombinant DNA technology to improve the ability of soil bacteria to "fix" nitrogen from the air. Soybeans and other legumes get their nitrogen from bacteria that inhabit nodules on their roots. Infection of soybean plants with super nitrogen-fixing strains could increase crop yields. Another future goal is to build nitrogen-fixing bacteria that will grow on the roots of corn and other crops that are unable to fix nitrogen. This remarkable accomplishment would reduce reliance on expensive commercial sources of nitrogen.

There is evidence, too, that the root cells of plants can be chemically altered so that they are more attractive to certain fungi that grow on or near the roots. In some way these fungi assist plants to increase the uptake of phosphorous from the soil. Scientists would like to develop crop cultivars that encourage beneficial fungi or bacteria to thrive in the vicinity of the plant roots and enable the plants to take up more phosphorous or nitrogen from the soil. Less commercial fertilizer would then be required to maintain or increase crop yields, and crop production costs would be decreased.

Additionally, some of the beneficial fungi that grow on, in, or near roots of certain crop plants make the roots less attractive to root-feeding nematodes that injure or deplete the root system with subsequent decrease in yields. Perhaps superior strains of these beneficial fungi can be identified, isolated, and increased and their spores used to inoculate soils of problem fields. Exudates from the roots of crop plants would stimulate the fungus spores to germinate and produce a web of protective mycelium around and in the root system. Less injury from harmful nematodes would result and yields would be increased without the cost of additional commercial fertilizer.

The possibilities of forming new kinds of plants and animals are endless. Man is now acquiring the knowledge needed to direct or control the evolution of living things, including human beings. However, before genesplicing can be accomplished routinely, scientists must learn much more about gene structure and what genes do. For one thing, very few genetic traits are controlled by one gene. Scientists believe 15 genes together control the ability of soil bacteria to absorb nitrogen from the air and transform it to a usable form in the roots of legumes. Probably more than 50 genes contribute to the process that makes corn protein different from soybean or tobacco protein. Despite these problems some scientists expect to one day transfer traits that cannot be transferred through crossbreeding through use of recombinant DNA.

Bacterial Vectors of DNA

The pathogenic bacterium *Agrobacterium tumefaciens* (Smith & Townsend) Conn., which causes crown-gall disease, attacks many species of plants. It is closely related to the nitrogen-fixing bacteria that grow in nodules on the roots of legume plants. In causing disease, *A. tumefaciens* enters susceptible plants through tiny wounds or injuries. DNA carried in the tumor-inducing (Ti) *plasmid* of the bacteria is transferred to the plant cells, by means of the juices bathing the wounded cells, transforming them into tumor cells. These affected cells are greatly stimulated to produce not only the many substances required for host cell division and enlargement, but also compounds that enable the bacteria to reproduce. The host cells begin to grow and divide rapidly forming soft galls or tumors on any wounded shoot or root; the galls may become many times larger than the root or shoot that bears them. The disease progresses slowly, stunting the plant but rarely killing it.

An important aspect of crown-gall disease is that plasmid DNA of the bacterial pathogen is transferred to the plant cell and is actually incorporated into the host's chromosomal DNA. This bacterial DNA then multiplies and becomes part of tumor tissue, free of bacteria.

There are many strains of the crown-gall bacterium, some of which are only weakly pathogenic. They can invade wounded cells and reproduce, but form only a few small galls or none at all. Apparently they do little harm to the infected plant. Realizing this, scientists have ingeniously used A. *tumefaciens* as a possible vehicle to insert foreign DNA (genes) into plants. Cultures of A. *tumefaciens* are treated with a restriction enzyme that "unlocks" or opens the ring of DNA (Fig. 8.3) in the chromosomal material contained in the plasmid of the bacterial cell. Solutions containing new, desirable DNA, for example the DNA that controls *phaseolin* (a protein) production in beans, are then added to the bacterial cultures; bits of the new DNA are absorbed into the bacterial chromosomes where it is replicated. Cultures of the harmless bacteria containing the new desirable DNA (gene) are then inoculated into experimental sunflower plants. As the bacteria begin to grow, the desirable DNA is transferred to the sunflower cytoplasm where it becomes part of the normal functioning process of the plant cells, imparting increased protein production to the sunflower.

Scientists can also isolate protoplasts from small tumors of sunflower tissue, regenerate them, and test the new plantlets for increased protein production. Active research is also underway using plasmid DNA of non-pathogenic strains of the crown-gall bacterium to insert factors controlling resistance to diseases into commercial crops.

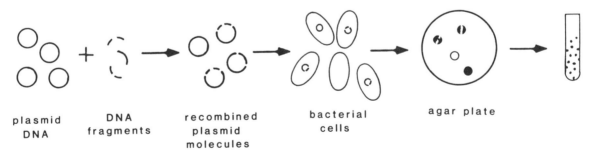

plasmid DNA   DNA fragments   recombined plasmid molecules   bacterial cells   agar plate

Fig. 8.3. Transforming bacteria by inserting foreign DNA into a plasmid DNA molecule. Plasmid DNA is treated with a restriction enzyme that opens the plasmid ring. Solutions of desirable DNA fragments (e.g., the DNA, obtained from a bean plant, that controls production of a useful protein) are added to the bacterial cultures. A few bacterial cells absorb the donor DNA into their plasmids. These cells are grown on selective media and yield a colony of identical cells or clones. Once the "new" colony has been found, it can be grown in quantity to produce the useful protein.
Courtesy American Scientist

Viral Vectors of DNA

It is safe to say that all, or almost all, organisms are attacked by viruses (Chapter 17). A plant virus is an infectious agent composed of ribonucleic acid (RNA) wrapped in a protein coat. Viruses can replicate only in living cells. In nature, virus particles are commonly injected into healthy plant cells through tiny wounds made by insects that pick up the virus while feeding on an infected plant. Once inside the cell, the protein coat of the virus dissolves, and the virus particle begins to make copies of itself. Many strains of viruses exist. Some cause severe symptoms on the host plant, others cause only mild symptoms.

Scientists are exploring the possibility of inserting beneficial, disease-resistance genes into the RNA of mild or harmless strains of plant viruses. These genetically altered viruses can then be injected into plants by an insect vector or with microneedles. Once inside the cell the virus particle carrying the beneficial gene or group of genes would begin to multiply and make the plant disease resistant.

9

Nematodes

Agriculture began when some early nomad thousands of years ago, gathered seeds from primitive cereal plants and instead of eating all the seeds, scattered some kernels by accident on the ground near a lake or stream. The next year when the tribe returned to the same spot, they were amazed to see the plants thriving. So, they reaped the harvest and planted more seeds. They no longer had to wander from field to field to gather food. They could reside in one place—in green pastures—to grow their crops and graze their cattle. Ever since that nomad unknowingly sowed the beginnings of agriculture, a potential enemy has lurked in every acre of farmland and garden soil. It is called the nematode.

General Nature and Importance

There are thousands of kinds of nematodes. Probably every living thing—animal or plant—bigger than a nematode is attacked or parasitized by them. The elephant, the gnat, the whale, the minnow, man, cattle, crop plants, birds, insects, other nematodes—all carry one to many kinds of these animals within their bodies.

Nematodes that feed on animals, however, will not or cannot feed on plants; likewise nematodes that feed on plants do not attack animals. This is a general rule in biology: Plant parasites do not attack animals; animal parasites will not attack plants.

Plant pathogenic species of nematodes are of tremendous importance as disease agents. All plants are parasitized by nematodes, but only a few hundred species of nematodes are known to be plant parasites. Full recognition of the amount of crop damage resulting from nematode diseases and the development of effective management measures have occurred only since World War II.

There are many reasons why scientists overlooked the importance of nematodes as a cause of reduced crop yields. Usually, farmers judge the health of plants on the basis of aboveground symptoms; rarely do they examine the roots of unthrifty plants. Moreover, the stunting, yellowing, wilting, and poor growth associated with nematode infections do not cause the concern that outright death of plants do. Most plant parasitic nematodes cause slow decline in yield from root damage rather than the easily seen damage caused by low fertility, acid soils, drought, or frost. Often the terms "tired land" or "sick soils" are used to describe fields on which plants make poor growth despite proper fertilization, weed control, and good cultivation. The land is not "tired"; it is often infested with plant-parasitic nematodes. Some plant-parasitic nematodes attack seeds (e.g., wheat gall nematodes), foliage (e.g., foliar nematodes), and even the water conducting tissues above ground (e.g., pine wood nematodes).

Other reasons these remarkable creatures escape notice are that they are microscopic in size or barely visible to the naked eye, and almost transparent. They spend their lives hidden in the tissues of host plants or in the soil near the roots. Another unusual thing about plant-parasitic nematodes is that they neither feed nor reproduce except on living plants. To survive, each individual nematode, after hatching, must reach a plant on which it can feed before its reserve food supply is exhausted. That means the nematode must swim through the soil in a film of water in search of food.

Although accurate information on the losses caused by nematodes is difficult to obtain, some agricultural scientists estimate the combined loss for all crops to be at least 10% of total agricultural production. This means an annual loss of several billion dollars worldwide. But more than that it means that many people who now go hungry could be fed, especially in developing countries.

Nematodes are a more serious and complex problem in many developing countries than they are in the developed countries. Most of the developing countries are in tropical or subtropical regions where the climate is very suitable for nematode activity and reproduction throughout the year. Warm, sandy soils are favorable to nematode infection, especially in irrigated areas used for continuous crop production. Perennial crops, and annual crops grown in the same fields year after year, are often so seriously attacked by nematodes that such crops can barely survive. Yet, the people in these developing countries must depend on perennial crops and nonrotated crops for their food supply.

Size, Distribution, and Anatomy

A nematode is a tiny threadlike animal resembling but not closely related to an earthworm. The nematodes as a group are second only to the insects in their total number. The name nematode comes from a Greek word *nematos,* meaning thread. Most nematodes range from 0.5 mm ($\frac{1}{50}$ in.) to 6 mm ($\frac{1}{4}$ in.) in length, but some that live in animals grow much larger. Approximately 80 adult lesion nematodes could lie side by side inside this hyphen: -. About 15,000 lesion nematodes could lie on one thumbnail without overlapping (see Fig. 2.1).

Nematodes are found everywhere and in almost everything: the top of the highest mountain, the bottom of the lowest valley, the most arid desert, and the most fertile farmland. They are found in the icy saltiness of Arctic seas and in the hot springs of Yellowstone Park. Many different kinds live in the soil, but most of these are harmless and some are distinctly beneficial. A single hectare of cultivated soil may contain hundreds of millions; as many as 100,000 nematodes have been recovered from a liter of garden soil.

The basic anatomy of a nematode is diagrammed in Fig. 9.1 and summarized below:

A Nematode Has	A Nematode Does Not Have
Digestive organs and glands	Eyes, ears, or a nose
Reproductive structures (male and female)	Blood vessels or blood
	Hair
Excretory structures	A skeleton
Muscles	A liver, gall bladder, or lungs
Nerves	Appendages (arms or legs)
Tough cuticle (nematode skin)	

A nematode moves by levering its body up and down, swimming in soil or surface water like an eel. A nematode moving with its side against the ground would closely resemble a snake gliding across sand.

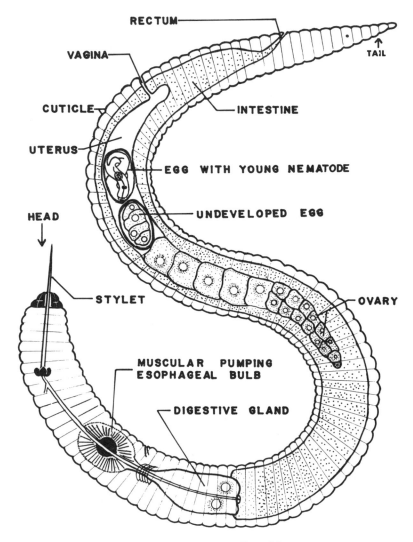

Fig. 9.1. **Anatomy of a plant-parasitic nematode (female).**
Courtesy R. P. Esser, Florida Dept. Agric.

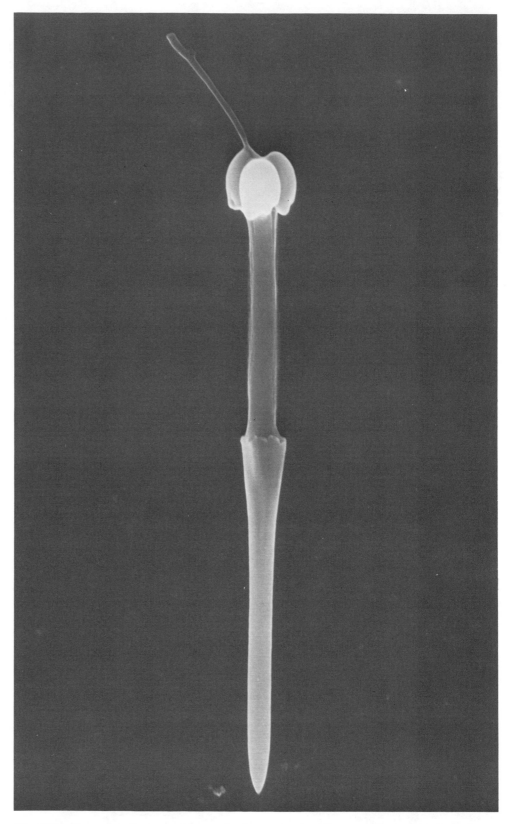

Fig. 9.2. **Scanning electron micrograph of a nematode stylet magnified about 5220 times.**
Courtesy J. Eisenback

Feeding Behavior and Plant Injury

Nematodes that feed on plants have mouths equipped with a *stylet,* which looks like a miniature hypodermic needle (Figs. 9.2 and 9.3). A nematode feeds on a plant by puncturing the cell wall with this stylet and sucking out the contents of the cell. Suction is induced by contractions of a muscular bulb in the esophagus of the nematode. Not only do nematodes suck out juices from plant cells, they also inject liquids into the cells and in this way may introduce viruses into plants. In addition, the tiny wounds or punctures in the tissues provide entry for bacteria or fungi.

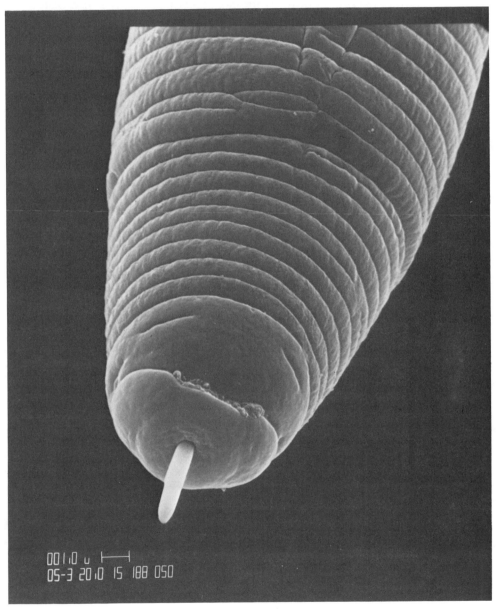

Fig. 9.3. **Scanning electron micrograph of the head of a nematode with protruding stylet magnified about 7840 times.**
Courtesy J. Eisenback and H. H. Hirschmann

The injection of the stylet into the cell of a plant may have drastic effects. Sometimes the cells are killed quickly. In other instances (e.g., root-knot nematodes), the cells remain alive and are stimulated to extra growth by the hormones and enzymes contained in the liquids injected by the nematodes. The adjoining cells increase in size and contribute to the supply of liquids continually sucked out by the feeding nematodes.

Some nematode species (e.g., the citrus nematode) attack only a very few kinds of plants, whereas others (e.g., the root-knot nematode) feed on many different kinds.

The various kinds of nematodes cause damage that interferes with the growth of plants. The weakened roots cannot efficiently absorb water and mineral nutrients from the soil. Reduction in the size of the root system by rotting or galling restricts its efficiency in obtaining the food and water the plant must get from the soil. Plants whose roots are attacked by nematodes lose vigor and become unhealthy. They may show symptoms of nutrient deficiency such as yellowing, stunting, or burning of foliage at tips and along edges. In addition, wilting occurs during hot parts of the day and the plants are less resistant to drought. Nematodes that attack seed or foliage result in poor seed quality or unsightly and diseased leaves.

On the basis of their feeding habits, nematodes are divided into two main groups:

1. *Endoparasites* enter plant tissues and feed from inside. Some move within the tissues and force their way between cells, injuring the tissue as they feed. Others remain in one spot, feed on only a few cells and become completely encased by host tissue.
2. *Ectoparasites* do not enter the plant. They feed from the outside and rarely penetrate the host tissues.

Life History, Reproduction, and Dispersal

Nematode eggs are found in great numbers in the soil. When the growing season begins, soil water carries substances exuded from plant roots to the dormant eggs causing them to hatch. From each egg a larva hatches which is capable of wriggling short distances in soil water. The newly hatched larvae, attracted to the plant roots, swim to the roots, enter, and proceed to feed and grow to adults. This usually takes place in less than a month if the soil is warm and moist.

The new adults lay more eggs. Several generations of nematodes may appear during the growing season, constantly producing new nematodes to feed on the roots and drain away energy from the plant.

In some species of nematodes the males are seldom seen. They are not necessary for reproduction and the eggs laid by females are usually not fertilized by males. The process by which reproduction occurs without males is called *parthenogenesis*.

Each female nematode lays many eggs inside or outside the plant. The young nematodes, called larvae, develop inside these eggs that have a tough shell which protects the young enclosed nematode from unfavorable conditions. The larvae come out of the eggs and develop into adults through a series of molts during which the outer cuticle is shed. The eggs may lie dormant in fallow land, sometimes for several years. Then, when the land is planted to a crop suitable for nematodes, the eggs hatch and new populations of nematodes build up in the soil. Each succeeding year, if conditions are favorable, many more eggs are present, spreading the nematode population in ever-widening circles.

Nematodes spread very slowly by their own efforts. Most nematodes never move more than a few centimeters from the spot where they hatched. Man himself has been the principal agent responsible for the spread of nematodes. Nematodes can be unseen contaminants on the parts of plants: corms, rhizomes, roots, tubers, seeds, flowers, leaves, and stems. Nematodes in the soil stick to tools, vehicles, and the feet of animals. Drainage and irrigation water also carry the eggs and larvae long distances to begin new centers of infection. In dry weather, strong winds may blow the eggs or nematodes from one field to another.

Enemies of Nematodes

All living things have their enemies. Just because a parasite feeds off a living host, this does not preclude the parasite in turn from being attacked by other parasites and predators.

Nowhere does "survival of the fittest" operate more cruelly than in the soil—this complex microcosm, this battleground of the living and the dying; of liquid, gas, and solid; of vegetable, animal, and mineral; of cycles within cycles; of surge and countersurge; of parasite on parasite; of poison juices, traps and nooses, predators, and foes. Plant-parasitic nematodes have many enemies in the soil. At any stage of their life cycle they may be captured and devoured by some other soil organisms. Indeed, certain types of nematodes eat other nematodes. Some of these predatory free-living nematodes have grasping or cutting teeth or both; others have long hollow stylets that are stabbed into the victim and suck out the body sap quickly.

Certain fungi also prey on nematodes. Some of these have sticky surfaces to which the nematodes adhere; others have loops or rings that close when a nematode starts to crawl through (Fig. 9.4). In either instance, the fungus grows into the nematode body and kills it.

Even though many plant-parasitic nematodes are killed by soil enemies, the whole population is seldom exterminated. Like other parasites, the nematodes manage to reproduce just a little faster than they can be wiped out. A single female root-knot nematode can lay more than 500 eggs. If only a few of them survive and reproduce in turn, an outstanding increase in population can take place during a warm growing season.

Fig. 9.4. Nematode-trapping fungi. Some fungi that live in the soil form loops or rings (arrow) that close when an unsuspecting nematode tries to crawl through the noose.
Courtesy M. K. Beute

Environmental Effects on Nematode Survival

Nematode populations do not remain static. They decrease when conditions are unfavorable and increase when conditions are favorable. The eggs can survive for a year or more in soil. Nematodes have been known to stay alive in soil flooded for 22 months. Species found in cold climates can easily live over winter and are not killed by freezing of the soil. Some species are killed if subjected to drying, but others enter a dormant state from which they can revive in a short time if moistened. A nematode that parasitizes rye has been revived after dry storage of 39 years.

In warm, damp soils, nematodes are very active. As the temperature cools, nematodes become less active. In warm climates, nematodes multiply at the rate of 5 to 10 generations a year and are also more vigorous in finding and attacking plants than in colder areas. Enormous populations result from this rapid buildup, which helps explain why nematode damage is so severe in warm regions.

The combined effects of sunlight and drying will kill the eggs of many nematodes in 30 min. However, the larvae are the weakest link in the nematode life cycle. The larvae are more active in moist soils, under conditions that favor rapid plant growth, but usually cause more damage during dry weather. However, they die quickly when exposed to direct sunlight, excessive heat or cold, or lack of moisture. As long as the moisture content in the soil is favorable to the growth of plants, nematodes prosper and multiply.

Nematodes and Other Diseases

Frequently, plants suffer from more than one disease at the same time, resulting in increased damage to or death of the plant. A plant may be able to resist the attack of one parasite; but when two or more attack the same plant, it may be overwhelmed. Nematodes are notorious partners in plant disease complexes. Plants attacked by both nematodes and other pathogens (bacteria, fungi, or viruses) often suffer severe injury, resulting in drastically reduced yields. Nematode-infested plants also may be quite susceptible to chemical injury from improper use of pesticides.

There are four major ways that nematodes team up with other pathogens to cause increased plant damage:

1. *Wounding Agents*—The wounds and punctures made by nematodes provide entry points for other pathogens. Nematodes open the doors and make it easy for other pathogens to enter. Once fungi or bacteria get inside, they thrive in the tunnels made by the nematodes.

2. *Host Modifiers*—When nematodes feed, not only do they suck fluids out of the host cells, they also inject digestive juices into the cells. These substances, including both enzymes and hormones, change the cells, increase their nutrient content, and make them a more suitable food source not only for the nematodes but also for other pathogens.

3. *Rhizosphere Modifiers*—Many fungi exist in the soil near living roots (the *rhizosphere*) as dormant spores that require a nutrient source before they can germinate and infect roots. Chemicals that leak out of roots are a principal source of nutrients. Nematodes affect this source by (1) making wounds that let increasing amounts of root liquids leak out and by (2) changing the composition of cell sap so that it is a more suitable food source for attacking fungi. The added food source not only allows the fungal spores to germinate, but it also supports growth of the fungi before they enter the root. The nematodes prepare the way for bacterial or fungal infections that would not otherwise occur.

4. *Vectors*—The nematode's stylet is a tiny, hollow spear that can pierce the host cell, suck out cell sap, and also inject salivary secretions into the cell. The stylet diameter of all plant-parasitic nematodes is less than 1 μm. However, the infective particles of plant viruses are smaller than 1 μm; thus, they can be carried in or on the stylet.

Bacterial and fungal populations sometimes occur in hugh numbers in the rhizosphere. Spores of these microorganisms may stick to a nematode's body and be transported by the nematode as it swims through the soil or feeds on plant roots. Since most spores have no means of self-transport, being carried about by nematodes could increase their chance of contacting a host.

Management

Nematodes are so widespread that it is practically impossible to eliminate them from infested fields. We will never get rid of the nematodes completely. We will never win the war against them. Therefore, the grand strategy in the war against nematodes is to learn how to live with them. The aim of management methods must be to keep the total nematode population to a tolerable level and to prevent the buildup of any one kind to an amount that could do great damage. No single method can effectively achieve these goals. Moreover, efficient nematode management requires a continuing year-round program aimed at keeping the crop as healthy as possible from the time the seed is planted until the crop remains are destroyed after the harvest. A combination of practices must be used if crops are to be grown profitably. Moreover, nematode management methods must fit into the overall IPM system in use on any given farm. Finally, any management program must be profitable. That is, the increase in the monetary value of the crop must offset the cost of the management measures.

Crop Rotation

Crop rotation is the most important cultural practice used in nematode management. Nematode management by crop rotation is based on the fact that parasitic nematodes can live and reproduce only when they can feed on suitable crop plants. In the absence of such plants, nematodes will simply starve to death. Therefore, in a crop rotation, susceptible crops are alternated with immune or highly resistant ones. While these are growing, the nematode population decreases. At the end of one or more growing seasons, depending on the kind of nematode present, the nematode population will have decreased to a point where the susceptible crop can be grown again with little damage. After one growing season, a new cycle of resistant crops must again be started. Rotations should be designed so that crops in any given field vary as widely as possible. Crops should not be planted in rigid sequence. Rather, the farmer should attempt to "rotate the rotation."

Small-scale farmers who raise poultry, swine, or cattle can use the animals as part of a rotation for managing nematodes (Fig. 9.5). In any given year the animals are restricted to a single portion of a fenced-in plot. The other sections are planted with various crops that are part of a 3-year rotation—crops

Fig. 9.5. Crop–animal rotation for managing nematodes.

Poultry House		
Year	Year	Year
1 Poultry	1 Maize	1 Garden
2 Garden	2 Poultry	2 Maize
3 Maize	3 Garden	3 Poultry
4 Poultry	4 Maize	4 Garden

Fig. 9.6. Chemical control of soybean cyst nematode, a relative of the root-knot nematode. The soybeans on the right are growing in soil that was fumigated (treated) with a nematicide 2 weeks before planting. The three rows on the left are growing in unfumigated (untreated) soil.

Courtesy D. P. Schmitt

for 2 years, animals for 1 year. The animal traffic prevents plants from growing, thus depriving nematodes of a host and the nematode populations decrease to a low level. The animal manure enriches the soil.

Chemical Management

Chemicals used for nematode control are called *nematicides.* Comparison of crops on treated and untreated soils demonstrates the effectiveness of nematicides (Fig. 9.6).

Nematicides have been in large-scale use since World War II, and their use has been increasing every year. It is not possible to kill all the nematodes in the soil even with increasing doses of these chemicals. Therefore, recommended application rates are now based on a cost/benefit ratio calculated to give the largest dollar increase in crop value for each dollar invested in the treatment. Usually, farmers expect to receive $3–$4 for each dollar expended. Nematicides are expensive; therefore, their use is usually justified only on high-value crops such as cotton, golf greens, greenhouse soil, lawns, nurseries, orchards, ornamentals, peanuts, seed beds, strawberries, sweet potatoes, tobacco, and vegetables. (Chemical management of nematodes is discussed in detail in Chapter 7, Soil Treatment section.)

Nematode-Free Transplants

For crops grown from transplants (annual flowers, cabbage, pepper, tobacco, tomato), the seed should be sown in seedbeds that have been previously fumigated with a nematicide that also kills bacteria, fungi, insects and weeds. Care should also be taken to avoid contaminating beds with infested soil while transplants are being grown.

Other Nematode Management Practices

Disinfecting Plant Materials

Nematode-infected plant material can be disinfected by dipping bare-rooted transplants, corms, or seedlings in solutions or emulsions or nematicides for a few minutes to an hour. The roots are soaked in the prepared solution for the required time, drained, and planted immediately.

Caution: Remember most nematicides are very poisonous and dangerous. Be sure no liquid touches the skin or clothing. Do not breathe the fumes.

Plowing After Harvest

As soon as possible after a crop is harvested, the roots should be plowed up. This exposes the roots and soil to the drying action of the wind and sun. It also prevents further reproduction of nematodes that would occur if the plants were allowed to continue growth. This destruction of the debris and remains of the preceding crop also helps manage insect pests and other pathogens that would continue to grow on a crop that is left standing in the field.

After harvest, if labor is available and economical, the plants may be pulled by hand, placed in stacks, allowed to dry, and then burned. This can also be done in home gardens or other small areas where only a few plants must be destroyed.

Using Resistant Cultivars

Considerable progress has been made in breeding cultivars resistant to certain nematodes. To be of value, a cultivar must be resistant to the nematode species and race prevalent in the region concerned and, in addition, have acceptable agronomic qualities, yield well, produce food or fiber of good quality, and, in general, be more profitable for the farmer than cultivars previously grown.

Some nematode-resistant cultivars have been bred and are widely used. However, the job of breeding resistant cultivars never ceases because new races of nematodes that will attack formerly resistant cultivars are continually evolving.

Nematode Assays

Several states now have a nematode advisory section, usually a subdivision of the state department of agriculture. Residents may send in soil samples to these laboratories and have the soil assayed for nematodes. On the basis of the analysis, suggestions are made as to what management measures are necessary. Nematode assays generally are just one of several services offered, including soil fertility and plant tissue analyses and a plant disease and insect clinic. Reliable services of this kind are not only a valuable asset to farmers, but also a necessity for efficient agriculture.

10

Diseases Caused by Nematodes

The representative nematode diseases discussed in this chapter (Table 10.1) are common and can be recognized fairly easily. It should be remembered that it is easy to mistake nematode diseases for those caused by some other organisms, and vice versa. Merely finding nematodes in diseased plant tissue or the soil is not conclusive evidence that they are the cause of the trouble. Nonparasitic types of nematodes often are found in great numbers in decaying plants, and the soil always contains a variety of free-living nematodes. Positive identification should always be obtained before starting expensive or troublesome management measures. On the other hand, nematodes should always be considered as a possible cause of plant diseases when root systems are galled, shortened, or reduced by rotting; the stems are shortened and thickened, and the leaves do not grow normally; or some other abnormal growth is noted.

Table 10.1. *Some Important Diseases Caused by Nematodes*

Disease	Pathogen	Principal hosts
Lesion	*Pratylenchus* spp.	Many crops, trees, and weeds
Root knot	*Meloidogyne* spp.	Most commercial crops; more than 3000 hosts reported
Stem and bulb	*Ditylenchus dipsaci* (Kühn) Filipjev	Alfalfa, corn, small grains, other crops
Stubby root	*Trichodorus* spp. and *Paratrichodorus* spp.	Alfalfa, cabbage, corn, cotton, grape, grasses, tomato, and others

Root-Knot Nematodes

Root knot is one of the most important plant diseases in the world. This widespread disease has changed human history by forcing the migration of peoples from one land to another. Root-knot nematodes have caused or contributed to famines. In many warm regions of the world, as the soil became more and more infested and yields continued to decline, farmers have had to move to other fields. When the new fields became heavily infested, farmers had to move again.

Throughout the world, root knot causes an average annual yield loss of about 5%. This loss would not be of great significance if it were evenly distributed, but it is not. Greatest losses occur to those who can least afford it, namely, the small farmers of underdeveloped countries. Their losses may be as much as 25–50% over wide areas of the available farmland.

In the southeastern United States, root knot is a primary threat to many crops. Root-knot nematodes must be managed if crops are to be grown profitably. Crop losses from root knot are heavy. In combination with other pathogens the disease can be disastrous.

Indirect losses associated with root knot are caused by the following:

- Secondary attacks by other pathogens
- Increased cost of weed control because of lack of competition by the dwarfed crops
- Loss of usefulness of cultivars resistant to other diseases but susceptible to root knot
- Inefficient utilization of fertilizer and water
- Cost of chemical treatment
- Loss of use of desirable fields because of the necessity for crop rotation

Root-knot nematode management is a good example of the value of agricultural research. Since World War II, nematologists and plant pathologists have demonstrated the importance of root knot in crop production and developed suitable management measures. However, the battle is never won. Research must continue in order to find new and more efficient ways to manage these ever-present parasites.

The root-knot nematode parasitizes more than 3000 plant species. Members of all the main crop families are attacked. Germinating seeds and young seedlings are particularly susceptible to infection. This includes the seedlings of some plants that are otherwise resistant.

Symptoms

Plants suffering from root-knot nematodes are stunted and yellowed. Severely infected plants may occasionally be killed, especially during dry weather. Within the same field, some areas may be severely affected, whereas plants in other areas may be free of symptoms. Indeed, individually affected plants may occur scattered throughout a field. It is unusual to find all plants in a given field affected. Plants with root knot may have symptoms of nutrient deficiencies or drought injury even in the presence of adequate fertilizer and water. During dry periods, affected plants develop severe symptoms. These plants become stunted and they wilt during the hot part of the day. The plants recover at night and wilt again the next day. Older leaves yellow prematurely, and the tips and edges "burn" or turn brown.

The most distinctive symptoms caused by root-knot nematodes are the *galls* or *knots* on the roots (Figs. 10.1, 10.2 and 10.9D). The galls vary in size from a pin head to compound galls more than 2.5 cm in diameter. The galls are irregular, spherical, or spindle shaped. Although the knots may be scattered on any part of the main root or its branches, they are most often found on tender rootlets like beads on a string. Small galls contain at least one

Fig. 10.1. Root-knot galls on bean roots about one-half natural size. Healthy roots should be long and fibrous without large galls.

nematode, whereas larger galls contain numerous females in all stages of development. Galls may be formed on tubers and on parts of the stem in contact with the ground.

Not only are galled roots rough and clublike, the number of feeder roots is decreased. Therefore, the roots can absorb less water and nutrients from the

Fig. 10.2. Root-knot galls on peanut shells (*left*) compared with slightly damaged shells (*right*).
Courtesy A. L. Taylor and J. N. Sasser

soil. This fact helps explain the wilting and nutrient-deficiency symptoms. Another conspicuous symptom is the premature rotting and death of the galled tissue, particularly late in the growing season.

Certain insects and nitrogen-fixing bacteria also cause galls on roots. However, positive identification of root-knot galls is made by breaking them open and looking for the pear-shaped, pearly-white females about as big around as the shank of a pin. They are large enough to see with the naked eye and easy to see with a magnifying glass, particularly in a portion of the root that has begun to rot so that the white females contrast with the brown root tissue. The egg masses of the root-knot nematode are also easy to see. These are brown, often as large as the female itself, and cling to the sides of the roots. When the sac containing the egg mass is lifted off, the female will be found embedded in the root.

Growers who suspect that they may have a nematode problem on their crops should send soil samples to a nematode assay laboratory, which usually charges a small fee. The local county agent can give instructions about how to collect and mail soil samples. Many agricultural consultants also offer this service as part of their "package" of agricultural advice.

Causal Agent and Disease Cycle

Root knot is caused by species of *Meloidogyne* [from the Greek *melos* (gourd or melon) + *oid* (resembling) + *gyne* (woman or female) = melon-like female]. There are at least 36 species, but the four most widespread that do the most damage, in order of their importance, are *M. incognita* (Kofoid and White) Chitwood, *M. javanica* (Treub) Chitwood, *M. hapla* Chitwood, and *M. arenaria* (Neal) Chitwood. The white, adult females are 0.4–1.3 mm long by 0.27–0.75 mm wide. The stylets of mature females range from 12 to 16 μm long. After the female becomes full grown, she begins to produce eggs in a yellow-brown, jelly-like sac fastened to the rear end of the female.

Each female forms on the average 400–500 oval or cylindrical, colorless eggs about 80 μm × 40 μm (Fig. 10.3). The eggs are found in great numbers in the soil. From each egg, a long, slender, shoestring-like larva, averaging 400 μm × 15 μm in size, hatches. The larvae can move or swim short distances in soil water. They are attracted to growing roots by the chemicals that leak out of them. Once a larva touches the root of a susceptible plant, it pushes, forces, and wriggles to the central part of the root containing the water-conducting tissue. Here under favorable conditions, the larva becomes a female and remains for the rest of her life. Reproduction occurs parthenogenetically in root-knot nematodes. Usually only a few males are seen, but when food is scarce or high populations of larvae are present, greater numbers of males develop.

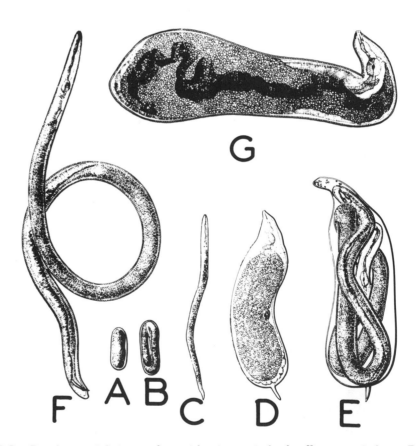

Fig.10.3. Developmental stages of a root-knot nematode. A—Unsegmented egg. B—Egg containing larva. C—Migratory larva as it occurs free in soil. D—Sausage-shaped larva as it lives in the plant root. E—Larval molt containing fully developed male. F—Adult male. G—Young female. All magnified 100 times.
After G. Steiner

Fig. 10.4. Adult female of the root-knot nematode with attached egg mass embedded in a protective jelly. Magnified 100 times.
After G. Steiner

Root-knot nematodes are called sedentary endoparasites because they remain stationary and feed inside the root. A nematode feeds on three to five cells only, puncturing these cells with the stylet and injecting chemicals, enzymes, and hormones that stimulate the plant cells to grow into *giant cells* that remain in a perpetual juvenile state. The enriched cell contents provide an excellent food source for the nematode, which begins to swell rapidly, first becoming a sausage-shaped and then a pear-shaped, whitish body. The plant cells surrounding the nematode are stimulated to divide forming the galls. After the female becomes full grown, she begins to produce eggs (Fig. 10.4). When the root decays, the galls rot, and the nematode eggs are released in the soil.

Environmental Effects

Root-knot nematodes are quickly killed in areas where the soil freezes, and infection is rare at 10°–12°C. Above 25°C root knot is severe, and the nematode's life cycle is completed in about 20 days. In warm climates, nematodes multiply at the rate of 5 to 10 generations a year and are also more active in finding and attacking plants than in colder regions.

As long as the moisture content of soil is favorable for the growth of the plants, soil moisture seems to play only a small part in development of root knot. Lighter, sandy soils are generally regarded as most favorable for increase of the nematode. These soils have large pore spaces that favor nematode movement and support fewer organisms that feed on nematodes.

Root Knot and Other Diseases

Although the root-knot nematode acting alone can cause serious damage to plants, it often teams up with other pathogens. When two or more pathogens act together, they may cause greater damage than the sum of their individual efforts. Three well-known examples of pathogens acting in concert with the root-knot nematode are the fungi *Fusarium* and *Phytophthora* and the bacterial-wilt organism. Not only do the nematodes provide the wounds that permit the other pathogens to enter the roots, but they also modify the host cells to provide a more suitable food source for the fungi and bacteria. Under the combined attack of two or more pathogens, the host plant quickly dies, and great losses occur.

Management

The general procedures discussed in Chapter 9 (Management section) and Chapter 7 (Soil Treatment section) are effective in managing root-knot nematodes.

Lesion Nematodes

In contrast to root-knot nematodes, which stimulate host cells to grow and swell, lesion nematodes (sometimes called meadow nematodes because they were first found in meadows) kill the cells on which they feed. Lesion nematodes are one of the primary factors involved in the general root destruction of many plants. These nematodes occur all over the world and feed on many kinds of crops. Different species of lesion nematode have broad and overlapping host ranges. They do much damage to crops such as alfalfa, apple, carrot, cherry, corn, cotton, grasses, peach, potato, tomato, tobacco, and many ornamentals. However, the direct damage caused by these pests is only part of the story. All root-invading nematodes provide sites for entry of secondary invaders and alter the host cells, thus causing compound damage. It is difficult to assess the damage caused by lesion nematodes as the disease is often associated with and leads to increases in other root diseases.

Symptoms

In a given field, lesion nematodes usually occur in well-defined areas, which vary greatly in size and shape. Diseased plants may occur individually or in small groups. Plants attacked early in the season grow poorly and become worthless. Aboveground symptoms resemble those caused by root-knot nematodes; that is, the plants are stunted and yellowed, and wilt during the hot part of the day. The edges of the leaves turn yellow, then brown and die ("rim fire").

Lesion nematodes do not form galls on the roots. The nematode larvae feed in the *cortex* of roots and kill the cells on which they feed. Fungi and bacteria then colonize the dead tissue. In the early stages the only visible symptom is a small red-brown lesion on the root. As the nematodes feed, the lesion enlarges, often girdles the root, and eventually severs it. Affected roots often shrivel and die; extensive root pruning results; feeding roots are destroyed, and the entire root system is reduced (Fig. 10.5). Diseased roots show various degrees of mutilation. There may be small yellow to brown to black lesions on the feeder roots. The lesions break open, and the cortex sloughs off like a sleeve, leaving only the vascular cylinder.

Fig. 10.5. Roots of plants grown in a field containing high lesion nematode populations. A—Soil untreated, most of the feeder roots have been destroyed by nematodes. B—Soil fumigated with methyl bromide, a luxuriant root system with numerous fibrous roots.
Courtesy C. J. Nusbaum

Fig. 10.6. **Lesion nematode eggs, larvae and adults in a feeder root about 0.8 mm in diameter.**
Courtesy C. J. Nusbaum

Often secondary roots appear above the nematode lesions, resulting in the formation of "blind" roots and a stubby, discolored root system. Surviving roots are grouped near the soil surface. Other nematodes and soil organisms also cause the same sort of damage. Positive identification of the disease, therefore, requires identifying the lesion nematode whose eggs, larvae, and adults are often seen in the cortex of infected roots.

Causal Agent and Disease Cycle

Lesion nematodes belong to the genus *Pratylenchus. Pratylenchus brachyurus* Godfrey, *P. scribneri* Steiner, and *P. zeae* Graham are important species. Different species are adapted to different crops and climates. Some thrive in warm, tropical regions; others do more damage in cooler regions. Lesion nematodes (Fig. 10.6) are stout and cylindrical, with a blunt head, strong short stylet, and rounded tail. They range from 0.4 to 0.7 mm long and 20 to 25 μm in diameter. In contrast to the root-knot nematodes some lesion-nematode species reproduce sexually. The eggs are laid singly, each gravid female averaging one egg a day for about 30 days. Eggs deposited in root tissue hatch in 6–17 days depending on environmental conditions. These endoparasitic, migratory nematodes wriggle through the cortex of infected roots, feed on, and destroy the cells. Eggs, larvae, and adults may overwinter in soil and in live or dead roots.

Management

Lesion nematodes can be managed by the procedures discussed in Chapter 9 (Management section) and Chapter 7 (Soil Treatment section).

Stem and Bulb Nematodes

Although it occurs worldwide, the stem and bulb nematode is most destructive in temperate climates. It is a serious pest of many crops including alfalfa, bean, clover, corn, flower bulbs, onion, pea, small grains, strawberry, and sugar beets. More than 400 species of plants are hosts to this nematode, which causes heavy losses by killing seedlings and dwarfing plants, making bulbs unfit for propagation, and reducing yields.

Symptoms

Retarded germination of seedlings and reduced stands often occur in fields infested with stem and bulb nematodes. Many seedlings are twisted, deformed, stunted, and yellowed. Swellings or galls develop on the shortened, thickened stems. The leaves become twisted, dwarfed, and distorted. Bulb tissue becomes soft and spongy and begins to decay. Nematodes can be found in large numbers in the affected parts but are difficult to see without a microscope.

Causal Agent and Disease Cycle

Ditylenchus dipsaci (Kühn) Filipjev, the stem and bulb nematode, is a long slender nematode (1.0–1.3 mm long; 30 μm in diameter) with a short stylet. The preadult larvae can withstand long periods of dormancy; they have been revived after storage in dry plant material for 23 years. During this dormant period they show no sign of life and their metabolic activity is practically at a standstill. They can withstand long periods of freezing or extreme drying, and with the return of suitable temperature (15°–20°C) and adequate moisture, the larvae become active again.

When the aboveground parts of plants are covered with a film of moisture, the nematodes move upwards to the new leaves and stems and enter through stomates or wounds. The nematodes feed on young succulent tissue and stimulate gall formation. Unlike root-knot nematodes, stem and bulb nematodes must mate to reproduce, and the females lay fertilized eggs. Eggs and larvae overwinter in dried, infected host material.

Disease incidence is most severe in cool climates in moist loam or clay soils. Damage is usually slight in dry years.

There are many pathogenic races of *D. dipsaci* which vary in their ability to infect different hosts.

Management

Since stem and bulb nematodes can stay alive in infested seed or bulbs for many months, the need for nematode-free seed is obvious. Use seed, bulbs, or other planting material from nematode-free plants. If bulbs from plants grown in infested soil are used, they can be freed of nematodes by heating in hot water at 46°C for 1 hr or soaking in a 0.5% formaldehyde solution at 43°C for 4 hr. Long rotations with resistant crops such as carrots, lettuce, spinach, and potatoes are helpful in reducing infestations. It helps to pull out and destroy diseased plants. Large fields can be treated with nematicides; small areas or greenhouse soils, fumigated or sterilized with steam (see Chapter 7, Soil Treatment section, and Chapter 6, Soil and Root Media Pastuerization section).

Stubby-Root Nematodes

Root-knot and lesion nematodes are classified as *endoparasitic* because they enter the roots and feed on the inside. In addition to these nematodes, there are a large number of *ectoparasitic* species, including the stubby-root nematodes. This group lives in the soil and feeds on the roots and other underground parts largely from the outside (Fig. 10.7). For many years the damage caused by stubby-root nematodes was unknown or overlooked because of their external feeding habits.

Fig. 10.7. Ectoparasitic nematode feeding on petunia rootlet from the outside. Note the collapsed cells punctured by the stylet.
Courtesy R. S. Pitcher

Unfortunately, it is hard to prove that a given ectoparasite is the cause of a certain disease. Even though a suspected nematode is found in large numbers in the soil surrounding roots of affected plants, many other nematodes may also be found, several of which may have caused the damage. Many of the ectoparasitic forms do not cause distinctive symptoms; any condition or agent that kills or injures the roots may cause similar symptoms. Furthermore, the life history of many of the ectoparasitic nematodes is not completely known; at certain times during the growing season they are found in large numbers in the soil surrounding the roots, but at other times the populations are low.

However, we now know that several genera of ectoparasitic nematodes are important plant parasites. These include the *dagger, lance, ring, stubby-root, spiral, sting,* and *stunt* nematodes. In addition to the damage caused by the feeding of these nematodes, some ectoparasites such as *Xiphinema,* the dagger nematode, also transmit plant viruses. If a nematode pierces the root cells of a virus-infected plant, some of the virus particles may be sucked up into the stylet. Then, if the same nematode moves to the root of a healthy plant nearby and injects salivary secretions containing virus particles into cells of a healthy root, it may become infected with the virus. The realization that nematodes can transmit plant viruses opens new avenues of investigation that may help in managing both nematodes and viruses.

Stubby-root nematodes are typical examples of ectoparasitic nematodes. They have been found in many parts of the world attacking a wide range of crops including beans, cabbage, clover, corn, grape, grasses, peach, tobacco, and tomato. They are found most frequently in sandy or porous soils. Stubby-root nematodes seldom kill the host plant, but they weaken the root system. Crops grown in fields with high populations of stubby-root nematodes cannot withstand drought conditions. The plants are stunted and produce reduced yields of low quality.

Symptoms

Poor stands and reduced growth are typical symptoms of stubby root early in the growing season (Fig. 10.8). Plants attacked by stubby-root nematodes remain stunted and *chlorotic* (yellowed). They have fewer and smaller leaves and branches than healthy plants. They wilt easily on sunny, hot days.

Stubby-root nematodes feed at or near the root tips. The injured roots are weakened and may stop growing. There is an abnormal growth of lateral roots and branch roots. These in turn are attacked by the nematodes and the result is a root system composed of short stubby roots often arranged in clusters

Fig. 10.8. **Stunted corn growing in soil with a high population of stubby-root nematodes. The farmer got no yield and it cost him money to prepare the land and plant the crop.** *Courtesy K. R. Barker*

(Fig. 10.9C). Parasitized roots are not killed, although they are usually darker in color than normal roots. The fine feeder roots typical of healthy root systems are conspicuously absent. Undoubtedly, such root systems cannot efficiently take up water and nutrients.

Causal Agent

Two closely related genera of nematodes, *Trichodorus* and *Paratrichodorus,* are called stubby-root nematodes. They vary in length from 0.5 to 1.5 mm and from 30 to 50 μm in thickness. These nematodes never enter the roots. They lay their eggs in the soil, and the life cycle is completed in about 20 days during favorable conditions. Both females and males are found. Populations vary considerably, depending on the kind and age of the plant, soil type, temperature, season of the year, and competition from other nematodes and pathogens.

Management

The general measures outlined in Chapter 9 (Management section) and Chapter 7 (Soil Treatment section) manage stubby-root nematodes. However, scientists need to know much more about these parasites before they can develop resistant cultivars and determine the most efficient nematicides to use on various crops.

Fig. 10.9. **Orchardgrass root systems. A—From nematode-free soil. B—From soil infested with lance nematode. C—From soil infested with stubby-root nematodes. D—From soil infested with root-knot nematodes.**
Courtesy N. E. McGlohon

11

Fungi

The fungi are a very old group of microorganisms, which have lived on earth for more than two billion years. Approximately 100,000 species of fungi have been described, and more are being found each year. Mushrooms, toadstools, molds, and mildews are fungi known by most everyone, but many fungi are so small that it takes an expert to find and see them through a microscope.

Structure and Importance

Fungi are microorganisms that contain no chlorophyll and usually reproduce by spores. They are sometimes classified as simple plants, but they have no roots, stems, or leaves. The body (*vegetative*) stage of a fungus consists of branching, delicate, threadlike structures called *hyphae*. Many hyphae together form an interlacing tangle known as a *mycelium*. The hyphae of many fungi are divided by crosswalls or *septae* into definite cells full of protoplasm. The protoplasm is held within a semipermeable cytoplasmic membrane containing one or more nuclei. Hyphae of other fungi possess no septae and are long tubes containing protoplasm; the streaming back and forth of the active liquid protoplasm is often visible. Mycelia may form on or inside the host and have different branching habits and structures, which help identify them.

Most fungi are beneficial. They are used for food, in medicine, and in industrial processes. Together with bacteria, fungi are also the principal agents of decay, and through the decomposition of organic matter they play an essential role in recycling of nutrients and the nutrition of plants. All fungi are capable of breaking down complex organic foods into simple substances and using these as sources of energy. Fungi decompose books, cloth, food, fruits, leather, meat, nuts, paper, stored seed, twine, vegetables, and wood.

Harmful fungi cause diseases of man, other animals, and especially plants. About 8000 species of fungi can cause plant diseases, and all plants are attacked by some kind of fungi. Some plant-parasitic fungi can attack many species of plants; others attack only one. Some fungi can grow and multiply only on a living host; these are called *obligate parasites*. Others can grow and multiply on dead organic matter as well as on living plants; these are called *facultative parasites*. In general, fungi that cause plant diseases will not attack humans, and fungi that attack animals also do not cause plant diseases.

Some substances produced by some fungi in food are toxic to animals. (e.g. aflatoxin).

Plant diseases caused by fungi increase the grocery bill of each family in the United States more than $600 each year. Hundreds of millions of dollars are spent annually for chemicals to combat pathogenic fungi. More millions are spent in breeding resistant plants; each state university employs numerous research scientists to discover new ways to reduce crop losses due to fungi. Our nation's forests lose more than 20 billion board ft of lumber each year to fungal and other diseases.

Growth and Reproduction

A hypha usually originates through the germination of a *spore.* The simplest spores are one-celled bodies of microscopic size, containing a nucleus and cytoplasm. They may or may not be able to swim. A spore germinates by pushing out, from a thin place in its wall, a tubelike or threadlike structure known as a *germ tube.* The germ tube grows into a hypha, which branches into many hyphae to form the mycelium. Growth is largely localized at the tips of the hyphae and is made possible by the absorption of liquid materials directly through the cell walls. The mycelium is usually substantial but hidden in the substance on which the fungus grows. The growing hyphal tips are able to penetrate directly through the cell walls of many plant tissues, even the hardest wood. They penetrate by means of organic substances known as *enzymes,* which they secrete. Enzymes have the power to dissolve or digest substances composing cell walls and other cell parts.

In some fungi the mycelium may pass into a dormant or resting stage by the formation of closely compacted hyphae, known as *sclerotia.* These vary in size from a few cells (*microsclerotia*) up to many thousands of cells. Some sclerotia may weigh up to 4 kg! However, most are quite small. Sclerotia also vary in color (colorless, pale yellow, brown to black) and shape (irregular, spherical, or elongate). In the form of sclerotia, fungi can hibernate or survive through conditions unfavorable for growth, such as cold, drought, heat, or the absence of a host. On the return of favorable conditions, the sclerotia may form hyphae that infect a host, form a new mycelium, or produce a reproductive body of some sort.

Many fungi reproduce by forming spores at the ends of, inside, or on specialized hyphae and structures. These spores are tiny, microscopic bodies the size of dust, and are produced in huge numbers; their function is similar to that of seeds in higher plants. They are carried by animals, diseased plants, insects, machinery, man, soil, wind, or water.

Fungi that live and grow in wet conditions often produce *zoospores* that can swim. These are produced in a *zoosporangium,* and after a brief swimming period, they germinate. However, most fungi produce nonmotile spores that are blown by the wind, splashed by rain, or moved with soil. These spores are formed in several different ways, and often a given fungus may produce more than one kind of spore. Many species have a spore stage in which they pass the winter or unfavorable conditions, and thus the spore becomes a *resting spore.* The different kinds of spores may be quite different in color, size, structure, and wall thickness.

To germinate, spores must be exposed to certain temperature and moisture conditions. Some spores require an actual film of water, whereas others are able to germinate in a moist atmosphere. Few spores germinate when it is very cold; the optimum temperature for germination of most spores is between 15° and 25°C.

There are four main groups of plant pathogenic fungi, differentiated for the most part by their hyphal characteristics and manner of producing spores: (1) *Phycomycetes* have mycelia without crosswalls (*aseptate*), may produce *sporangia* and *zoospores* (swimming spores) and sexual spores such as *zygospores* and *oospores.* (2) *Ascomycetes* have septate mycelia, may produce asexual spores called *conidia* and usually produce sexual *ascospores* in a saclike structure called an *ascus.* (3) The *Fungi Imperfecti* are a collection of fungi which have septate mycelia but usually produce only asexual spores in nature. (4) *Basidiomycetes* have septate mycelia, produce sexual *basidiospores,* in most cases, and may produce other spores such as *teliospores* and *uredospores.*

Distribution and Dispersal

Fungi grow almost everywhere. They thrive in the soil, in living and dead plants and animals, and on all kinds of organic matter. They grow both in fresh and salt water; the dustlike spores may be blown thousands of kilometers. Fungi can use many substances for food. Provided sufficient moisture is present, they will grow on most organic matter. Some even grow on paint or oil, others attack some synthetic materials and insulating material.

In general, fungal plant diseases can be classified into one of two types: those caused by soilborne fungi and those caused by airborne fungi. As the names imply, the means by which each type of pathogen is spread differ. In some cases, this difference is reflected in how a disease develops and what management methods are appropriate. Several diseases caused by soilborne fungi are discussed in detail in Chapter 12; those caused by airborne fungi are considered in Chapter 13.

Although the greatest attention is usually paid to diseases of crop plants, fungal diseases of shade and forest trees also cause severe losses. The management of such tree diseases is particularly difficult because the host plants are widely dispersed and live for many years. Several important fungal diseases of trees and decay in wood are discussed in Chapter 14.

12

Diseases Caused by Soilborne Fungi

Soilborne fungi cause some of the most widespread and serious plant diseases. Spores or mycelium of many of these fungi can overwinter or survive adverse conditions in soil or on plant debris, and once an area has become infested with soilborne fungi, it is generally difficult to get rid of them. The representative diseases discussed in this chapter are summarized in Table 12.1.

Phytophthora Root Rots

Among the fungi called "water molds," because they thrive in wet soils, are the *Phytophthora* spp., which occur almost everywhere crops are grown. This genus contains several species of soil-inhabiting fungi that cause serious, widespread, and difficult-to-control root and stalk rots in warmer regions.

Black shank of tobacco is prevalent throughout the southeastern United States and is one of the most important diseases of tobacco. It spreads rapidly, and a large part of the crop can be destroyed in a short time. Black

Table 12.1. Some Important Diseases Caused by Soilborne Fungi

Disease	Pathogen	Hosts
Root rots	*Phytophthora* spp.	Tobacco, soybean, ornamentals, and others
Damping-off	*Pythium* spp.	Many crops
Root rots, leaf blight, and damping-off	*Rhizoctonia solani*	Beans, tomato, cereal grains, turfgrasses, and many others
Root and stem rots	*Fusarium* spp.	Many crops
Southern blight	*Sclerotium rolfsii*	Many vegetable crops, peanut, soybean, forage crops, and others
Crown and stem rots, watery soft rot	*Sclerotinia* spp.	Many vegetable crops, forage legumes, peanut, soybean and others
Black rot	*Cylindrocladium crotalariae*	Peanut, soybean, other legumes

shank reduces yields so drastically that it is impossible to make a profit unless the disease is managed adequately. Another *Phytophthora* sp. causes an important root rot of soybeans, which can greatly reduce soybean yields. Other *Phytophthora* spp. attack many ornamental plants including azalea, camellia, deodar cedar, heather, mountain laurel, rhododendron, and yew. A *Phytophthora* sp. causes little leaf of pine, the most important disease of shortleaf pine in the Piedmont region of the South (see discussion in Chapter 14). Red stele of strawberry, the most serious disease attacking this crop, is also caused by a *Phythophthora* species.

Black Shank

Symptoms

Black shank causes a damping-off of young seedlings in seedbeds. When vigorously growing plants 30 cm or more high are attacked, the first indication of the disease is the sudden wilting or drooping of the leaves. After a few days the leaves begin to turn yellow and hang down the stalk. The entire root system and the base of the stalk turn black, decay, and die. On older, taller plants, the stem may turn black 30 cm or more above ground. Soon the leaves turn from yellow to brown and shrivel; in a few days the plant is dead. When the stem of a diseased plant is split in half lengthwise through the black lesion, the pith appears dry and brown to black and is usually separated into platelike disks (Fig. 12.1). This is one of the most characteristic symptoms of this disease.

Fig. 12.1. Black-shank infected tobacco plant with stalk split lengthwise to show typical decay, "disking" of pith, and dead root system.

Causal Agent and Disease Cycle

This disease is caused by the fungus *Phytophthora parasitica* var. *nicotianae* Dast. (Breda de Haan) Tucker. Anyplace where plant roots are growing in waterlogged soils provides good conditions for *Phytophthora* spp. to attack. The hyphae of *Phytophthora* spp. do not have crosswalls and they branch profusely. In water the fungus produces tiny, colorless, lemon-shaped *sporangia* (about 25 μm × 35 μm), each fastened to a fragile, slender stalk called a *sporangiophore* (Fig. 12.2). When the sporangia are ripe and temperature and moisture suitable, the protoplasm inside each sporangium divides into about 30 colorless, bean-shaped zoospores. These remarkable spores can swim (most spores cannot) by propelling themselves with two whiplike structures (*flagella*). Zoospores are attracted to roots by the leachates or plant juices that leak out of the roots. They attach themselves to the root surface, and in a few minutes a germ tube starts to develop. Within 2 hr the germ tube has penetrated the root and infection has begun. The mycelium penetrates all through the root tissues killing them.

Fig. 12.2. Phytophthora parasitica **magnified about 400 times. A—Sporangia. B—Zoospores emerging from sporangium. C—Stages in zoospore germination. D—Chlamydospore germinating to form sporangium. E—Chlamydospores. F—Hyphae. G—Oospores with antheridia.**
Courtesy P. D. Dukes

When the food supply is all used up or the root tissues begin to die, the hyphae swell and form a type of *resting spore* called a *chlamydospore*. As the roots decay, the chlamydospores are released into the soil. These spherical spores, about 25 μm in diameter, can survive in soil for 5 years or longer. They enable the fungus to survive during drought, cold, and lack of food. When the soil becomes warm and moist and a young rootlet grows near a chlamydospore, it is stimulated to form sporangia with zoospores. *Phytophthora* spp. also form sexual spores called *oospores*. These are about the same size as chlamydospores and can be confused with them. New pathogenic strains of *Phytophthora* are produced when oospores germinate.

Black shank is a warm-weather disease. The fungus thrives in warm, wet soils. When temperatures range from 20° to 35°C with frequent rains, resulting in high soil moisture, zoospores are produced in great numbers. These are washed and/or swim from one place to another infecting any tobacco plants they happen to contact.

Management

To effectively manage black shank, farmers must coordinate several practices, all of which are necessary. The backbone of any management program is the use of resistant cultivars. Black-shank-resistant cultivars are available that are also resistant to root knot and bacterial wilt. Without these, other management measures are of little value.

All farmers in an area infested with black shank should try to grow their own disease-free transplants and avoid seedbed failures. Seedbed soil should be fumigated to kill bacteria, fungi, nematodes, and weeds (see Chapter 7, Soil Treatment section). Plant beds should not be watered from streams or ponds that drain infested fields.

Root knot and black shank each increase the damage done by the other. Therefore, root knot also must be managed. Fields that are to be planted in tobacco should be treated with appropriate multipurpose chemicals that kill nematodes, fungi, and bacteria.

Rotations with immune crops (corn, cotton, grasses, peanuts, small grains, sweet potatoes) are also valuable in managing black shank by helping to reduce the amount of fungus in the soil. The same rotations should not be used year after year. Change the cropping sequence, i.e., "rotate the rotation" to prevent root-knot nematodes or other pathogens from building up to dangerous levels.

After harvest, stalks should be cut and the roots plowed to expose them to the drying action of sun and wind. This reduces carryover of nematodes to the next crop.

Phytophthora Root Rot of Soybean

Symptoms

Phytophthora root rot can occur on soybeans of any age. If the fungus attacks seedlings early, damping-off may result. If seedlings are slightly older, root and stem rot can cause wilting, yellowing, and seedling death. Older plants are killed gradually and plant vigor is reduced throughout the season. Upper leaves turn yellow and plants wilt. Affected plants die, but the withered leaves often remain attached for a week or more. Taproots turn dark brown and this discoloration may extend up to the fourth or fifth node. Inside the plant the cortex and vascular tissues are also darkened.

Causal Agent and Disease Cycle

The causal agent of this disease is *Phytophthora megasperma* (Drechs.) var. *sojae* A. A. Hildebrand. The characteristics of the fungus and the general disease cycle are similar to those for *P. parasitica* var. *nicotianae*.

Management

Several physiological races of *P. megasperma* var. *sojae* can attack soybeans. Thus, it is important to know which fungal races are present in an area. Once this is known, an appropriate high-yielding Phytophthora-resistant cultivar can be selected from the numerous ones available. Additionally, growers should always plant in warm (18°C or more), well-drained, fertile soil. Rotation with nonhost crops such as wheat or corn helps reduce the amount of fungus in the soil.

Phytophthora Root Rot of Ornamentals

Symptoms

The symptoms of Phytophthora root rot on azalea vary with the cultivar. On Kurume types, new leaves are smaller than normal with interveinal chlorosis, purple discoloration, and defoliation. This chlorosis is often confused with a deficiency of iron or other nutrients. Applications of iron and fertilizer usually improve the green color of the leaves only for a short time. Excessive yellowing and loss of older leaves are typical symptoms on Snow Azalea. Usually, plants slowly decline in vigor and die branch by branch over a period of several months to years, but sometimes they can die rapidly. Roots are dark colored, brittle, and often limited to the upper part of a container of soil. The reddish-brown discoloration advances to the larger roots and eventually to the main stem.

On rhododendron the primary symptoms of Phytophthora root rot are a rapid wilting and death of leaves (Fig. 12.3). Leaves droop but remain attached to the limb. Root symptoms are similar to those on azalea.

On *Camellia japonica* the main symptoms are a gradual decline in vigor and loss of older leaves. Root symptoms are similar to those on azalea. *Camellia sasanqua* is tolerant to the pathogen. Yew trees die rather suddenly, with the foliage turning reddish brown, when infected by *Phytophthora* species. Roots are also reddish brown, and the discoloration may extend into the main stem. Fraser fir is also susceptible.

Causal Agent and Disease Cycle

The pathogen which causes this disease is usually *Phytophthora cinnamomi* Rands. The fungal characteristics are similar to those of *P. parasitica* var. *nicotianae*.

Fig. 12.3. The rhododendron plant on the right with drooping, rolled leaves is dying from *Phytophthora* root rot. The roots of the apparently healthy plant on the left intertwine with those of the diseased plant. Therefore, the healthy plant will probably become infected and die also.

Courtesy R. K. Jones

Phytophthora root rot of ornamentals is favored by high soil moisture and warm soil temperatures. The frequency of root rot increases in areas where surface water collects, and in shallow soils with underlying rock or compacted hard pans. Heavy clay soils; overwatering; tight, poorly drained soil mixes in containers; and containers with a few small holes in the bottom are also predisposing factors that favor disease development. Settling of containers into sawdust or other similar material used as a ground cover for weed control may also increase the disease.

Management

The only practical management of Phytophthora root rot in nurseries involves avoiding the disease, either by the production and use of disease-free plants or by the prevention of disease development on established plants.

Producing Disease-Free Plants. To assure proper sanitation of planting areas, the following practices are recommended:

• Remove all old rooting media, old roots, and plant debris from propagating benches and beds and the surrounding area.

• Treat all unpainted wood, flats, frames, and benches with 2% copper naphthenate. Treatment with a 1:10 dilution of a commercial chlorine bleach or fumigation with recommended chemicals under plastic will eradicate the fungus from wood surfaces, pots, tools, and equipment.

• Remove all plants from the greenhouse at the beginning of the propagating season and disinfect the interior.

• Fumigate rooting media or potting media under a plastic tarp or with steam (82°C for 30 min) before use. Effective steam treatment requires that all the soil reach 82°C, so the actual treatment time will be longer. Store rooting and potting media on higher ground than the rest of the nursery to prevent its contamination by surface runoff water from irrigation and rain. A paved area is suitable for this purpose.

Except in areas with deep, well-drained, sandy soils, it is not possible to grow disease-free, Phytophthora-susceptible plants in the field. Planting on raised beds with improved drainage will reduce the symptoms and nursery losses due to Phytophthora root rot, but the plants will likely have the fungus in the soil or roots and die after transplanting. Planting in raised beds with bark incorporated in the planting site is important, however, in landscape plantings to prevent root-rot development.

Much cleaner plants can be grown in pathogen-free, well-drained potting mixes in containers. This potting mix should contain little or no peat moss or soil and no sawdust. If soil (do not use heavy clay soil) is used in the mix, it should be treated with steam, methyl bromide, or a mixture of methyl bromide and chloropicrin before use. Potting mixtures containing large amounts of bark (without many fine particles) and sand have a high air volume (20–25%) and good water drainage; such mixtures are not favorable to the growth and development of *Phytophthora* species. Containers should be placed on a well-drained site on crushed rock, gravel, or coarse bark so that water can drain away from the container and surface water contaminated with the fungus cannot enter the drainage holes in the bottom of the container.

Do not overwater plants susceptible to Phytophthora root rot. Even in well-drained potting mixes, *Phytophthora* species can develop if constantly overwatered. All *Phytophthora*-susceptible plants should be grouped together in a nursery so that they can be given the special attention necessary to keep them free of disease. If at all possible, water *Phytophthora*-susceptible plants with well water. Lake, pond, or stream water can be contaminated with spores of *Phytophthora* species by surface runoff water. Drainage ditches should be constructed so that surface water does not drain into irrigation water sources other than wells. Also, do not overfertilize or set plants too deep since these practices can increase the chance of *Phytophthora* root rot development.

Preventing Root Rot on Established Plants. Phytophthora root rot on established ornamentals must be prevented because few chemicals are effective in controlling this disease after aboveground symptoms become obvious. The following suggestions may aid in the prevention of root rot:

• Purchase disease-free plants from a reputable nursery. Avoid plants that lack normal green color, appear wilted in the morning, and evergreen plants that have excessive defoliation or dark discolored roots.

• Plant *Phytophthora*-susceptible plants in well-drained areas. If excess water from any source collects in the planting site, avoid planting susceptible plants. If soil is heavy clay or does not have good internal drainage, set susceptible plants in raised beds and thoroughly mix a porous material such as bark (not sawdust or peat) into the bed. The material should be incorporated to a depth of 20–30 cm. In some areas drain tile plus gravel, placed 15–30 cm below the surface, may also help reduce excessive soil moisture.

• Do not set a new plant any deeper than the soil level in the container or the soil line in the nursery.

• In areas where susceptible plants have died, replant with plants that are not susceptible to root rot.

Damping-Off

Damping-off disease of seedlings occurs all over the world. Indeed, the soil in every cultivated field probably is inhabited by fungi that cause this disease. Greenhouse and nursery growers are constantly threatened by outbreaks.

Losses from damping-off vary with the crop, season, soil type, and weather. The disease is responsible for many seedbed failures and poor stands. Sometimes during wet, cold planting seasons, it is almost impossible to get a crop off to a good start because of damage from damping-off fungi. Damping-off is a common disease of flower and vegetable seedlings and causes extensive losses through death of seedlings or production of low-quality plants that perform poorly in the field.

Symptoms

Damping-off fungi may rot seeds soon after they are planted, or kill young seedlings before they emerge from the soil (Fig. 12.4). This is called *pre-*

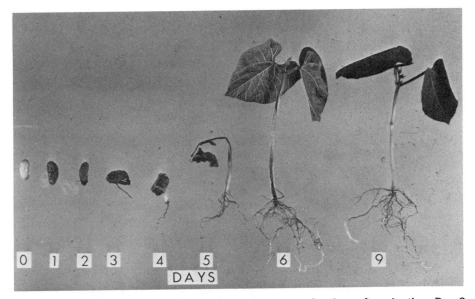

Fig. 12.4. *Pythium* damping-off of snap bean on successive days after planting. Day 0, a healthy seed; 1 and 2, seed rot; 3, preemergence damping-off; 4, 5, and 6, postemergence damping-off; 9, a healthy seedling.
Courtesy G. Abawi

Fig. 12.5. **Damping-off caused by** *Pythium.* **A—Damping-off of young seedlings. Note shrivelled stems and fallen plants. B—Damping-off of older seedlings. Under environmental conditions favorable to fungi, this circular area will grow progressively larger.**
Courtesy Conn. Agric. Exp. Stn.

emergence damping-off. Later these fungi may attack young seedlings above ground or near the soil line (*postemergence damping-off*). A brown, watery, soft-rot develops, the young stems are girdled, and the seedlings fall over (Fig. 12.5). In advanced stages of the disease, affected tissues collapse, and the stems either disintegrate or become shrivelled and limp. Affected areas in a seedbed, containing shrunken, collapsed, or dying seedlings, may be 30 cm or more in diameter.

Causal Agent

Many *Pythium* spp. cause damping-off diseases. These fungi live on dead plant and animal material as *saprophytes.* They also can act as low-grade parasites, which attack any young succulent tissue, particularly fibrous roots and tender stems. The slender, profusely branched mycelium grows rapidly and produces many sporangia, each one containing 30–100 zoospores which are capable of swimming in water. When the zoospores are released, they swim about in the soil water for a few minutes. If the roots of susceptible plants are growing nearby, the zoospores are attracted to the roots by exudates that serve as nutrients or stimulants. Once the zoospore touches a root, the zoospore stops swimming and within 2 hr produces a germ tube, which quickly penetrates the root tissue to start a new infection. Food from the root tissue enables the fungus mycelium to grow rapidly and produce a new crop of sporangia in a few days. The fungus also produces oospores or sexual spores. Since oospores require a resting period before they germinate, they are also called resting spores. Oospores germinate by germ tubes which develop into mycelium or by producing zoospores in sporangia. The type of germination of both sporangia and oospores is controlled by temperature. At temperatures above 18°C, germination is by germ tubes; temperatures between 10° and 18°C induce zoospore production.

Another fungus, *Rhizoctonia solani* Kühn also causes damping-off. This fungus causes other diseases in addition to damping-off and will be discussed in the next section.

Disease Cycle

The fungus enters seeds by direct penetration of the moistened, swollen seed coat, or through cracks or wounds. It is attracted by the chemicals leaking from seed and seedlings during germination. The fungus penetrates the embryo or emerging seedling tissues through mechanical pressure and enzymes, which dissolve those parts of the walls that hold the cells together. Further breakdown results as the fungus grows between and through cells. Other enzymes digest the proteins of the cell, and the fungus consumes the cell contents. Thus, infected seeds and tissues are killed and turn into a rotten mass consisting primarily of fungus and undigested host parts. The severity of damping-off depends upon several factors, including soil moisture and pH. *Pythium*-induced damping-off is most severe in wet soils and at high pH (Fig. 12.6).

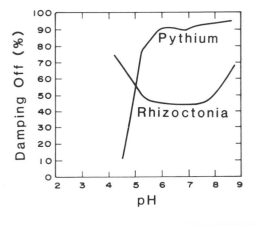

Fig. 12.6. Relation of soil moisture (*above*) and pH (*below*) to the incidence of damping-off caused by *Pythium* and *Rhizoctonia.*

Pythium spp. also can infect fleshy vegetable fruits or leaves in the field, in storage, in transit, or at the market. For instance, an entire cucumber fruit on wet soil may be rotten and soft 3 days after being infected by *Pythium* species.

Management

The best management approach to damping-off is to avoid it altogether. Once damping-off has started in a plant bed or seedling flat, it may be difficult to contain. Several general practices are employed to prevent damping-off:

- Proper soil preparation and management to provide good aeration, drainage, structure, and water-holding capacity; fertilization based on soil test results to provide adequate plant nutrition.
- Proper soil treatment to reduce the level of fungi that cause damping-off.
- The use of certified, fungicide-treated seed with high germination.
- Proper seeding rates to avoid dense plant stands, poor air movement, and low light intensity.
- Strict sanitation to avoid reinfesting treated soil with disease-causing organisms.

Many outbreaks of damping-off can be attributed to poor sanitation practices after treating the soil. For small bedding-plant operations, it may be practical to buy a sterilized soil mix and eliminate the soil preparation, fertilization, and treatment practices described above.

Once damping-off has started in a bed or flat, it may be managed by providing drier conditions for seedling growth. This can be done by ditching inside and outside the plant-bed structure to give efficient water drainage and by removing dirty covers, overhanging branches, or other obstructions that shade the bed to increase the amount of light.

If these preventive management measures fail, several fungicides are available that may be effective if applied as a drench or heavy spray as soon as the first symptoms of damping-off are observed. The fungicides in combination with proper growing conditions may stop the spread of damping-off. Rapid identification of the causal fungus should be obtained so that proper chemicals can be applied. Several applications of the fungicide may be necessary. Check the label carefully to be sure the proper fungicide and application rate is used on a particular crop.

Rhizoctonia Rots

The pathogen *Rhizoctonia solani* Kühn is one of the most important plant-pathogenic soil-inhabiting fungi. It causes more different types of diseases to a wider variety of plants than any other plant-pathogenic fungus known. It even parasitizes other fungi. This destructive fungus occurs worldwide in crop plants and also is common in forests and noncultivated desert land.

Rhizoctonia Diseases and Their Symptoms

One of the most common diseases caused by *R. solani* is seed rot. The fungus invades the planted seed from infested soil; the seed decays and never germinates. Often the rotted seed provides food for the fungus, which then grows through the soil and rots adjacent seed. In other instances the seed germinates, but the fungus attacks and destroys the seedling before it emerges from the ground. These seed rots and preemergence damping-off result in poor stands, particularly in seed flats and nursery beds. Any condition that delays germination may increase the chances of damping-off. Young seedlings are also attacked after they emerge from the soil. Young, succulent tissues are easily invaded, and decay of the young stem at the soil level causes the plant to fall over. Damping-off tends to occur in circular or irregular patches that represent foci of infection.

As plants mature, they become increasingly resistant to *R. solani*. Some species, however, including cabbage, forest tree seedlings, snap bean, and tomato, are attacked through the transplant stage. The dark brown discoloration of the stalk at or near the soil line, extending upward for several centimeters, is a characteristic symptom of sore shin or wire stem (Fig. 12.7). Affected stems are smaller than normal, discolored, tough, and woody. When diseased plants are set in the field during conditions favorable for disease, a poor stand often results.

Cereal grains in some countries sustain serious losses from root and crown rot caused by *R. solani*. Underground fleshy roots are often affected by cankers as well as by crown rot. Carrots, celery, potato tubers, radishes, sugar beets, tulip bulbs, and turnips become discolored and rotted by the fungus.

Fig. 12.7. Snap bean seedlings affected with sore shin, caused by *Rhizoctonia solani.* The seedling on the right is healthier than the others.
Courtesy G. Abawi

During warm, wet weather, *R. solani* may spread through the top of the plants; this is called web, leaf, or thread blight. The closer plants are to each other, the greater the spread of aerial blights. Beans, tall fescue, and other turfgrasses, lespedeza, rice, and soybeans are attacked this way. Brown patch of turfgrasses is an aerial blight (Fig. 12.8). The diseased areas, often 1–2 m in diameter, have a dark purple-green advancing margin in which the webbed mycelium is visible in the early mornings. At first the grass appears water-soaked; soon it turns dark; finally the leaves wilt, then die, and turn light brown. During hot, humid weather a smoky gray to black border of wilted, "webbed" grass called a smoke ring, may be seen in the early morning, bordering the light brown diseased area. The crowns and roots are rarely invaded. Leaves are infected through stomates and mowing wounds. Sclerotia are formed near the base of the plants. Because the grass grows close to the soil and when there is much dew, the high humidity enables the fungus to grow rapidly. "Poling" (using a long bamboo pole as a brush) the grass in the early morning to knock off the water droplets reduces the rate of fungus spread. *Rhizoctonia solani* that causes brown patch requires high temperatures and relative humidity to thrive and produce the disease symptoms.

Apparently, any aboveground plant part, especially fruits, that come into contact with soil infested with *R. solani* may become rotted. Infection occurs through wounds or the intact skin. Tomato fruits, either green or ripe, develop small, firm, brown spots where they touch the soil. These later enlarge, become soft, and have concentric zones of dark tissue. Mycelium and sclerotia may develop on the surface.

Belly rot of cucumber is caused by *R. solani*; similar spots occur on bean, eggplant, pea and peppers. Small infections at harvest time continue to develop in storage; infected seeds help start the disease in the next crop.

On Irish potatoes, one of the best-known symptoms of *R. solani* is the appearance of black sclerotia on potato tubers. It is called "dirt that will not wash off", black scurf, black speck, and black scab. The dirty, ugly black mycelium really does not reduce the value of the potatoes; it just makes them unattractive. However, if potatoes are stored under warm, wet conditions, the fungus will begin growing and the potatoes will rot.

Fig. 12.8.   **Brown patch on Bermudagrass. Some diseased areas may be 2 m (6 ft) or more in diameter.**

Fig. 12.9. Sixteen isolates of *R. solani,* growing on a special agar in a petri dish which vary in color, abundance of aerial mycelium, sclerotial formation, and pathogenicity.
Courtesy E. E. Butler

Causal Agent and Disease Cycle

Rhizoctonia-induced diseases may be confused with diseases caused by *Pythium* or *Phytophthora* species; however, the characteristic mycelium of *R. solani* readily distinguishes it. The hyphae are broad and coarse (8–12 μm in diameter), colorless when young, but turning dark brown with age. Young branches are inclined at angles of 45–90° from the direction of growth of the parent hyphae and are usually constricted at the point of origin. A septum or crosswall is always present near the base of the branch. Sometimes tan to black sclerotia ranging up to 6 mm in diameter form on decaying plants, and these help identify the pathogen.

There are many different strains of *R. solani* (Fig. 12.9). Some grow more

Fig. 12.10   Top: Infection hyphae of *R. solani* attached to the outside of a bean stem showing typical branching and cross-walls, magnified about 400 times. *Bottom:* Section of a bean stem cut through the center of dome-shaped infection cushion of *R. solani* showing numerous hyphae.

Courtesy T. Christou

rapidly above ground; others are adapted to life in the soil. Strains also differ in host range, pathogenicity, pH response, and temperature requirements.

R. solani is a primitive parasite with simple food requirements. It grows rapidly, and as a saprophyte it can live on dead or decaying plant debris for many months. It persists in the soil as hyphae or sclerotia, which may contain large amounts of stored food. Infections often start from sclerotia (Fig. 12.10), which can be carried on roots, seed, and tubers and in soil dry enough to blow as dust. Even in moist soil sclerotia can survive 5 years.

Exudates that leak out of germinating seedlings and actively growing roots stimulate the growth of *R. solani* from sclerotia. The sclerotia germinate by hyphal growth. Once contact with host tissue is made, the hyphae penetrate directly or form "infection cushions," composed of hyphal clusters on the susceptible tissues, which cause discoloration and death. The hyphae penetrate living cells through this dead tissue or invade directly through cracks and wounds. The invading hyphae grow rapidly through the host tissue causing it to turn brown and collapse. The fungus produces numerous enzymes that digest the host tissue and make it suitable for fungus use. Under favorable conditions, symptoms appear within 3 to 7 days after infection.

Diseases caused by *Rhizoctonia* species can occur in relatively dry soils and over wide pH and temperature ranges (see Fig. 12.6). They occur at cool temperatures in seedbeds and in the field when temperatures are high; they usually are most severe at temperatures that are unfavorable for the host. Excess solutes in the soil, resulting from overfertilization, increases the incidence of these diseases. Increased disease severity results from the interaction of *R. solani* on plants already infected with root-knot nematodes.

Management

Several basic guidelines, if followed, will aid in managing diseases caused by *Rhizoctonia* species.

- Do not use seed contaminated with sclerotia of *R. solani*.
- Do not use infected corms, roots, or tubers for planting stocks.
- Use certified seed treated with fungicides whenever possible.
- Use disease-free transplants grown in soil sterilized with steam, dry heat, or methyl bromide.
- Plant seed at the proper depth or practice shallow seeding so that seedlings emerge as quickly as possible.
- Plant at temperatures favorable for the host.
- Use resistant cultivars when available.

To manage brown patch of grasses, avoid seeding in hot weather and overfertilization with nitrogen; mow frequently to prevent matting of foliage and to permit more light penetration and aeration. Rake and remove clippings if there are heavy accumulations. Water deeply and infrequently instead of lightly at frequent intervals. Spray with recommended chemicals when the disease is first noticed and at regular intervals during warm, moist weather.

Fusarium Root Rots and Yellows

The genus *Fusarium* contains many species that attack a large number of crops causing seed rots, root rots, foot rots, stalk rots, wilts, yellows, ear and kernel rots. This variable fungus, composed of many pathogenic strains, lives in the soil and attacks all cultivated crops and many wild plants. The fungus is capable of attacking all parts of the plant. Great losses may result especially on susceptible cultivars when weather conditions are favorable to disease development.

Fusarium Diseases and Their Symptoms

Fusarium seed rots and root rots occur when spores carried on seed or present in soil infect germinating seedlings. Infected seed often rot and die before they emerge from the soil. Small lateral roots are usually killed soon after infection (Fig. 12.11).

Fig. 12.11. **Fusarium stem rot of snap beans.**
Courtesy G. Abawi

Fusarium stalk rot of corn is present each year and may cause considerable damage, particularly if abundant rainfall occurs during the latter part of the growing season. Stalks previously injured by cold, leaf diseases, or chinch bugs are especially susceptible to attack. Diseased stalks ripen prematurely and are subject to excessive stalk breaking. Stalk rots not only add to the cost of harvesting, but the ears on broken stalks may touch the ground, increasing their chances of rotting.

Ear and kernel rots of corn may cause serious losses when conditions are favorable for their development. Severe infection not only reduces yield but also lowers the quality and grade of the grain produced. In addition to these losses, the infected kernels may contain substances toxic to animals and are worthless as seed. The fungus typically causes a pink reddish rot of the ear. Apparently healthy looking kernels may be heavily contaminated with spores during harvesting or shelling. Spores carried on the seed will infect the new corn crop to repeat the infection cycle.

Fusarium root rot is a limiting factor in snap bean and pea production. It is particularly severe during cool, wet soil conditions. Depending on soil and environmental conditions, *Fusarium* spp. may act alone as a rotting agent; but more often they act in combination with species of *Pythium, Rhizoctonia,* and other soilborne pathogens. Often such plant disease interactions make the growing of legume crops unprofitable in certain years.

Other *Fusarium* spp. penetrate roots principally through wounds made by nematodes. The fungus grows through the water-conducting tissues, producing toxins that kill the cells, stunt the plant, and yellow the leaves. Often the plant wilts and dies from lack of water. Symptoms usually appear only on one side of the stem and progress upward until the foliage is killed and the stem dies. The water-conducting tissues just under the bark turn brown; this discoloration is visible in cross sections of stems near the base of infected plants. The fungus mycelium can be seen, with the aid of a microscope, growing in these tissues.

Causal Agent

Fusarium species produce three kinds of spores (Fig. 12.12). *Microconidia* are colorless, single-celled, spherical spores about 6–15 μm in length and 3–5 μm in diameter. The sickled-shaped, colorless *macroconidia* have 3 to 5 crosswalls and average 30–50 μm in length and 2–5 μm in diameter. The smooth, spherical, single cells called chlamydospores are produced on older mycelium and average about 10 μm in diameter. All three spore types are produced in soil or on infected plants. After infected plants die, the fungus and its spores are returned to the soil where they may persist indefinitely.

Contaminated seed, infested soil, and infected transplants help spread the pathogen from one place to another.

A

B

C

D

Fig. 12.12.   Fusarium oxysporum A,B,C, magnified about 1000 times. A—Macroconidia. B—Microconidia. C—Chlamydospores. D—Conidiophore, magnified about 380 times.
Drawing by H. Hirschmann after S. Johnson

Management

Seed treatment with the proper fungicides helps prevent seed rots and seedling diseases caused by *Fusarium* spp. *Fusarium*-free seedlings of transplant crops should be grown by treating seedbed soils with methyl bromide (see Chapter 7, Soil Treatment section). Resistant cultivars (cabbage, cantaloupe, tomato) should be used when available; rotation should be practiced, and root-knot nematodes controlled.

Southern Blight

Southern blight, also known as southern wilt and southern stem rot is a serious and frequent disease of many crops in the southeastern United States. It attacks a number of vegetable crops including bean, cantaloupe, carrot, pepper, potato, sweet potato, tomato, and watermelon. In addition, several field crops (cotton, peanut, soybean, sugar beets, tobacco) and forage crops (clover, alfalfa) are affected.

Southern blight occurs to some extent every year and at times causes severe crop losses. Losses are more severe in fields where the same crop has been grown for several years. Affected plants usually are scattered, but occasionally large areas ("hot spots") of diseased plants are found in a field. The fungus may also decay harvested produce, particularly carrots.

Symptoms

The disease is recognized by the wilting and yellowing of leaves; when the plant is pulled up, it is apparent that the lower stem and upper roots are infected. In watermelon, one or two runners may be affected; in cantaloupe, the melons are usually affected first. The edible roots of sweet potatoes have circular, sunken, dark gray surface spots 5–10 mm in diameter. The stems of erect plants such as green bean, pepper, potato, tobacco, and tomato are usually rotted at the soil line. The fungus causes a watery fruit rot on cantaloupe and tomato; in carrot, the whole root becomes decayed.

The rots are not associated with an offensive odor, at least initially. A white, moldy growth is evident on affected stem tissues and adjoining surface soil; later smooth, light tan to dark brown "mustard seed" sclerotia (about 2 mm in diameter) are evident in the mycelium. The sclerotia are diagnostic for the disease (Fig. 12.13).

Causal Agent and Disease Cycle

Southern blight is caused by the soilborne fungus *Sclerotium rolfsii* Sacc. The white mycelium grows in strands. The sclerotia begin to form in culture after about a week as small intertwined, white tufts of mycelium, but become yellow to brown with age. Although the sclerotia may survive 5 years in dry soil, they live for only about 2 years or less in moist soil. Once the food source is used up, the mycelium dies after a few weeks.

The southern-blight fungus overwinters as sclerotia and mycelia in host

Fig. 12.13. **Mustard-seed sclerotia of the southern-blight fungus forming on diseased soybean stem.**
Courtesy D. P. Schmitt

debris in the soil. The fungus is generally restricted to the upper 5–8 cm of soil and will not survive at deeper depths. In most soils the fungus does not survive in significant numbers when a host is absent for 2 years or more. The fungus is more active in warm, wet weather; it requires the presence of undecomposed crop residue to initiate infection.

Southern blight is favored by warm weather, high soil moisture, and aeration. Other factors influencing outbreaks are not fully understood. Both cropping history and cultivation practices during late summer, as well as climatic factors, influence disease severity. Outbreaks are usually more severe following excessive hilling or bedding of plants by cultivation or when there is an accumulation of fallen leaves around the base of plants. The fungus is capable of growing on undecomposed litter from a previous crop or other slowly decomposing organic matter, using this material as a food base. Once established and making vigorous saprophytic growth, the fungus can infect uninjured host tissue, although injuries make penetration easier.

Management

The control of southern blight is difficult, but losses can be reduced by following an integrated program of disease management over a period of several years.

Proper land preparation is vital in managing southern blight. The previous crop must be well decomposed before planting, and this may require disking the field several times in the fall and in the spring. The previous crop litter should be buried with a moldboard plow equipped with heavy-duty concave disk-type coulters to a depth of 20–30 cm; this plow depth must be below the depth of later cultivation equipment. The crop litter should be covered with 8–14 cm of soil. None of the buried litter should ever be brought back near the soil surface during the current season by cultivation.

Other important management practices include the following:

- Avoid problem fields.
- Rotate susceptible crops with corn, small grain, or other grass crops.
- Do not throw soil with debris against plant parts during the growing season.
- Control foliar diseases since dead leaves on the ground may trigger infection. Weeds should also be controlled early in the season for the same reason.
- Use a fungicide in the transplant water or as a drench spray for tomatoes and peppers. Dust formulations may also be applied over V trenches prior to transplanting. Follow all label directions carefully.

Crown and Stem Rots and Watery Soft Rots

Several species of the soilborne fungus *Sclerotinia* are causal agents for crown and stem rots and watery soft rots, which occur in more than 170 species of plants worldwide. Especially susceptible vegetables include bean, cabbage and other crucifers, carrot, cucumber, eggplant, Irish potato, lettuce, pepper, and squash. Many other crops such as alfalfa, clover, soybean, peanut, tobacco, and turfgrasses are susceptible. Some *Sclerotinia* species also attack ornamentals. On many vegetable crops, diseased plants wilt and flop in the field; but of equal importance is a watery soft rot that continues to damage the harvested vegetables during transit and storage.

Symptoms

On vegetable crops, diseases caused by *Sclerotinia* species can be recognized by a soft, watery rot with white, moldy growth on stems, petioles, and leaves (Fig. 12.14). Infection may start on leaves that touch the soil. The fungus grows through the petiole to the stem turning it black and often girdling it. Then the plant flops. If conditions remain moist, a large amount of cottony, moldy growth can be seen on the dead tissue. As this growth progresses, hard, black, irregularly shaped sclerotia form on the surface or in the

Fig. 12.14. Mycelium and sclerotia of *Sclerotinia* on diseased tobacco stem magnified about two times.
Courtesy M. Tirelli

pith of the stems. The sclerotia look like rat dung and are diagnostic signs for the disease. Sclerotia range from 2 to 10 mm in length, tend to be about two to three times longer than they are thick, and are white to pinkish inside. After the infection has apparently dried up, the line of demarcation between healthy and diseased tissue is very sharp. Often the diseased tissue is a light straw color.

On alfalfa, clovers, and other forage legumes, the disease is called crown and stem rot, but the fungus attacks all parts of the plant. Symptoms first appear in the fall as small, brown spots on leaves and petioles. The heavily infected leaves turn gray-brown, wither, and become covered with the white, cottony mycelium, which spreads to the crowns and roots. By late winter or early spring, the crowns and basal parts of infected stems show a brown soft rot that extends downward into the roots. Consequently part or all of the new growth of the infected plant wilts and dies. Diseased stolons become soft and rotten. Black sclerotia form on the white mycelium. On peanuts, symptoms are similar but the disease occurs in late spring and summer.

In newly seeded legume crops, crown and stem rots cause severe reduction in stands. The damage is often confused with winter killing. The disease occurs in patches throughout a field, but when conditions become favorable, the patches merge, and all plants in large areas of a field may be killed.

On closely mowed grasses such as turf grasses on golf greens, the disease (called dollar spot) appears as light yellow or tan circular spots 5–6 cm in diameter, about the size of a silver dollar, hence the name. As the disease progresses, these spots merge into large areas of dead, straw-colored grass.

Sclerotinia species also cause a petal blight or flower spot of azaleas and camellias. Infected petals turn brown and lifeless before they fall to the ground. Often the diseased flowers cling to the plant and present an unsightly appearance. Sclerotia form on the fallen petals beneath the plant.

Causal Agent and Disease Cycle

Sclerotinia spp. overwinter (or oversummer) in the soil as black sclerotia which measure 5–10 × 3–6 mm in diameter. Sclerotia can survive up to 7 years in dry soil. However, if the soil stays warm and moist, the sclerotia will be killed in less than a year. Following a dormant period, if moisture and temperature are suitable, the sclerotia germinate by resuming vegetative growth or by forming as many as 35 small (5-mm-diameter) mushroom-like bodies, called *apothecia* (Fig. 12.15). The apothecia produce enormous numbers of spores which are blown about and cause primary infections. They are the only infective spores in the disease cycle.

Once the fungus is established, it continues vegetative growth as long as there is sufficient moisture and host tissue available. The disease develops when moisture is high and the temperature cool. Temperatures between 15° and 20°C are ideal, but the disease will progress slowly at 3°C and at temperatures as high as 25°C. For the sclerotia to germinate and form apothecia, the temperature must be below 20°C; for spores to cause infection, the relative humidity must be above 90%.

Management

Management of diseases caused by *Sclerotinia* spp. is difficult. Clean cultivation, deep plowing to bury sclerotia, and long rotations are helpful. Spraying with fungicides may reduce the damage on field-grown vegetables and ornamentals. The most promising method of management for pasture legumes appears to be the breeding of resistant and/or adapted cultivars.

Fig. 12.15. Sclerotia and apothecia of *Sclerotinia,* natural size.
Courtesy D. E. Ellis

Watery soft rot can be controlled on plant materials in storage and transit by harvesting and shipping disease-free stock only. The disease can be held in check by precooling before loading for shipment and maintaining storage and transit temperatures as near 0°C as feasible.

Azalea and camellia petal blight can be managed by fungicidal sprays. Winter applications of fungicidal dusts or drenches to the soil underneath plants severely infected the preceding spring will help kill overwintering sclerotia.

Management of *Sclerotinia* species on greenhouse crops requires continuous efficient management. Greenhouse soils should be fumigated with methyl bromide or heated to kill the sclerotia. At the end of the season, all plants should be removed from greenhouses. Livestock manure and plant mulches should not be used unless disinfested by heat or methyl bromide. During the growing season, the surface soil should be kept as dry as possible and the air at as low a humidity as possible. Fungicides should be used as needed.

The most effective management of dollar spot on golf greens is to prevent outbreaks by maintaining healthy vigorous turf with proper cultural practices, such as the following:

- Providing adequate fertility based on soil test recommendations (nitrogen-deficient turf is more susceptible to dollar spot).
- Reducing thatch more than 2 cm thick by proper mowing.
- Maintaining adequate soil moisture. Avoid light, frequent watering; deep, infrequent watering is more effective during drought.
- Removing grass clippings where disease is active.
- Cleaning mowers and other equipment of infected grass before moving to uninfected areas.
- Using recommended fungicides. Repeated use of some chemicals may result in tolerant strains of the fungus; therefore, use chemicals only where necessary and alternate use of different fungicides.

Cylindrocladium Black Rot

Cylindrocladium black rot (CBR) of peanuts, caused by the fungus *Cylindrocladium crotalariae* (Loos) Bell & Sobers, was first identified on peanuts in Georgia in 1965. Since then it has spread to most of the peanut-growing states in the southeastern United States, where it now causes serious losses every year.

Black rot has also been found on soybeans in North Carolina. Symptoms appear as clumps of three or four dead plants in a pattern similar to that which occurs with southern blight. Examination of stems near the soil line, however, often reveals the characteristic reddish-orange fruiting bodies of *C. crotalariae,* called *perithecia.* Root systems are heavily rotted. The fungus also attacks other legumes, but does not appear to harm corn, cotton, or tobacco.

Other species of *Cylindrocladium* cause a serious damping-off and seedling blight of both hardwoods and conifers. Before the discovery of adequate management measures, Cylindrocladium blight seriously hampered the production of healthy forest tree seedlings.

Symptoms and Causal Agent

On peanut, CBR causes an overall stunting and yellowing of the plant and leads to death, in some cases, as the season advances. Previously healthy plants show signs of wilting on warm days and may eventually collapse. Infected plants have severely blackened and somewhat shredded tap roots almost completely stripped of lateral roots. The blackened area extends up to the crown of the plant and, in some cases, on the stems above the ground. Immature pods and pegs also show various stages of blackening. Within infected roots the fungus forms small, black, survival structures about the size of ground pepper. These structures, called *microsclerotia,* are released into the soil as the surrounding tissues decay and account for the ability of the fungus to survive in the soil for long periods (as least 3 years), even in the absence of susceptible host plants.

A certain diagnostic sign of black rot is the appearance of small, reddish-orange, spherical *perithecia* (0.4 mm in diameter). These reproductive structures of the fungus appear in dense clusters on stems, pegs, and pods just above or beneath the soil surface. They are especially abundant in moist areas under dense foliage. *Ascospores* are exuded from each perithecium as a viscous, yellow ooze, 2–3 weeks after the perithecium appears. These spores may play a role in the spread of the disease within a field but are not capable of long-distance movement under normal circumstances. Cylindrical colorless conidia (75 μm $\times$ 6 μm) are also produced abundantly and contribute to local spread of the pathogen.

Management

The most important means by which *C. crotalariae* spreads from farm to farm and within farms is by small microsclerotia (seed), which are present in diseased peanut roots and vines and are scattered in plant debris during digging and harvesting operations. Management of the disease is aimed primarily at curtailing the spread of the fungus in peanut fields.

In fields where only small localized areas of newly diseased plants occur, all affected plants should be removed from the field before harvest. In fields with larger infested areas, farmers have two options:

- Disking under all diseased plants before the digging and harvesting operation, making sure to avoid these areas with equipment during harvest.
- Digging out and removing all diseased plants where feasible and either burning them or burying them in a sanitary landfill.

Soil and plant trash should be removed from all equipment used in infested fields since the microsclerotia can be moved on it. Equipment should be washed with a strong stream of water immediately before it is moved from infested areas.

Small grain and corn are not susceptible to CBR and therefore do not increase its severity. Consequently, rotations of 3 to 5 years, using corn, small grain, or perennial grasses, such as fescue, are recommended for infested fields. In addition, cotton and tobacco are possible alternative crops on such land since the fungus will not increase in soils planted to them. As a preventive measure, farmers should plant resistant cultivars when they become available. Until additional effective means for disease management are found, growers are urged to avoid planting peanuts or soybeans in fields where CBR has occurred.

Effective management of the species causing Cylindrocladium seedling blight is possible by fumigating forest nursery seedbeds with methyl bromide. There is evidence, however, that strains of the fungus resistant to methyl bromide have developed. Therefore, in some areas it has been necessary to apply increasing amounts of the gas to nursery soils to eliminate the fungus so that disease-free seedlings can be produced.

13

Diseases Caused by Airborne Fungi

Because airborne fungi can be spread long distances on the wind, they can cause devastating losses, particularly on crops that are planted over large regions. In a given year, for example, the entire U.S. wheat crop from the Mexican border to the Canadian border may be exposed to black stem rust. These pathogens have the potential of causing, and in the past have caused, widespread epidemics on a variety of crops. Several of the most important diseases caused by airborne fungi are discussed in this chapter (Table 13.1).

Table 13.1. Some Important Diseases Caused by Airborne Fungi

Disease	Pathogen	Host(s)
Stem rust	*Puccinia graminis* f. sp. *tritici*	Wheat/barberry
Leaf rust	*Puccinia recondita*	Wheat
Cedar-apple rust	*Gymnosporangium juniperus–virginianum*	Cedar/apple
Corn smut	*Ustilago maydis*	Corn
Covered smut	*Tilletia* spp.	Barley, oats, wheat
Loose smut	*Ustilago nuda*	Barley, oats, wheat
Late blight	*Phytophthora infestans*	Potato, tomato
Downy mildew	*Plasmopora viticola*	Grape
Downy mildew	*Pseudoperonospora cubensis*	Cucurbits
Powdery mildew	*Erysiphe* spp. *Podosphaera* spp., *Spaerotheca* spp.	Many plants
Blast	*Pyricularia oryzae*	Rice
Gray mold	*Botrytis* spp.	Many crops
Leaf spot	*Cercospora* spp.	Bananas, beets, celery, forage legumes, pea- nuts, soybeans
Leaf blight	*Helminthosporium maydis*	Corn
Brown spot	*Helminthosporium oryzae*	Rice
Leaf spot	*Alternaria* spp.	Many crops
Anthracnose	*Colletotrichum* spp.	Many crops
Scab	*Venturia inaequalis*	Apple
Soft rot	*Rhizopus* spp.	Crops with fleshy organs

Rusts

Some people say that rusts are the most important plant diseases in the world. They occur everywhere, reducing the yields and quality of the important grain crops on which all human civilization rests.

Thousands of years ago when early man found he could use seeds of wheat like grasses as a reliable and palatable food source, he ceased to be a nomad. He no longer had to wander from place to place to find green pastures for his flocks; he became a tiller of the soil. His diet improved and he lived longer. Civilization changed as people began to cluster together in villages and cities. Jacob Bronowski says in his book *The Ascent of Man* that the single largest step in the ascent of man was the change from nomadic life to that of settled village agriculture. Yet, down through the centuries rusts have threatened man's principal food crops.

Wheat is the most important food crop in the world today, providing 20% of the world's food calories and occupying approximately 20% of the world's cultivated land. Most is grown in the Northern Hemisphere, principally in the United States, Canada, Europe, and Russia. Wheat is an annual grass that is adapted to clay-loam soils and semiarid climates. Most wheat is eaten as bread, but it also furnishes livestock feed and various products for brewing and other industries.

Wheat falls prey to several rust diseases (black stem rust, leaf rust, and stripe rust). Sometimes, annual crop losses to wheat rusts amount to more than one million metric tons worth billions of dollars. In years when wheat rust is severe, there is not enough grain to make bread. Famines result. So far in the United States we have never had a great famine. We have always had "our daily bread," and reports of starvation are seldom heard. But in other countries famine is a way of life.

In addition to the rusts of cereal grains, many other grasses are attacked by rusts, reducing their forage value. Cattle grazing on rusted grasses find them less palatable and sometimes develop allergies after eating infected forage.

Another rust, the cedar-apple rust, is an annual threat to apple crops. Many vegetables and ornamental crops also are infected by their own special rusts. Forest trees also are attacked. Fusiform rust of pine, one of the most important diseases of this valuable pulp and timber crop, is discussed in Chapter 14.

Black Stem Rust of Wheat

Symptoms and Disease Cycle

Wheat steam rust is caused by *Puccinia graminis* (Pers.) f. sp. *tritici* Eriks. & Henn. In the course of the disease cycle, five types of spores are produced. These are red spores (uredospores), black spores (teliospores), basidiospores, pycniospores, and yellow spores (aeciospores). Figure 13.1 diagrams the life cycle of black stem rust.

The first symptoms of stem rust on wheat are small yellow flecks that develop into long, narrow, yellow blisters on the stem, leaves, and leaf sheaths

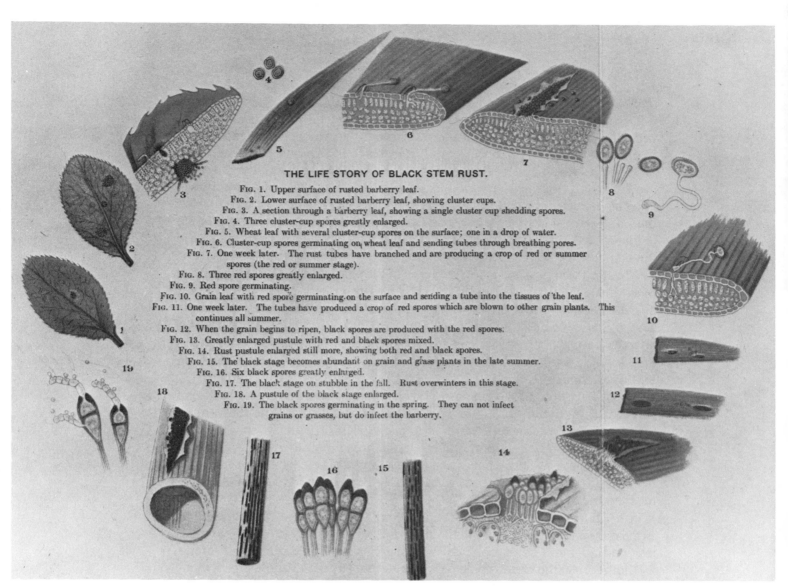

THE LIFE STORY OF BLACK STEM RUST.

Fig. 1. Upper surface of rusted barberry leaf.
Fig. 2. Lower surface of rusted barberry leaf, showing cluster cups.
Fig. 3. A section through a barberry leaf, showing a single cluster cup shedding spores.
Fig. 4. Three cluster-cup spores greatly enlarged.
Fig. 5. Wheat leaf with several cluster-cup spores on the surface; one in a drop of water.
Fig. 6. Cluster-cup spores germinating on wheat leaf and sending tubes through breathing pores.
Fig. 7. One week later. The rust tubes have branched and are producing a crop of red or summer spores (the red or summer stage).
Fig. 8. Three red spores greatly enlarged.
Fig. 9. Red spore germinating.
Fig. 10. Grain leaf with red spore germinating on the surface and sending a tube into the tissues of the leaf.
Fig. 11. One week later. The tubes have produced a crop of red spores which are blown to other grain plants. This continues all summer.
Fig. 12. When the grain begins to ripen, black spores are produced with the red spores.
Fig. 13. Greatly enlarged pustule with red and black spores mixed.
Fig. 14. Rust pustule enlarged still more, showing both red and black spores.
Fig. 15. The black stage becomes abundant on grain and grass plants in the late summer.
Fig. 16. Six black spores greatly enlarged.
Fig. 17. The black stage on stubble in the fall. Rust overwinters in this stage.
Fig. 18. A pustule of the black stage enlarged.
Fig. 19. The black spores germinating in the spring. They can not infect grains or grasses, but do infect the barberry.

Fig. 13.1. The life cycle of black stem rust of wheat published by E. C. Stakman in 1918.
Courtesy USDA

of young seedlings or plants at any stage of growth. Even the glumes and beards become infected. The blisters (*pustules*), 3–10 mm in size, break open revealing a powdery mass of brick-red or rust-colored spores. These are called summer spores, *uredospores,* or repeating spores. Each spore is about 25 μm long. A single pustule may produce more than 350,000 spores. The red spores escape from the pustules and are blown by the wind to other plants where they germinate, invade the plant, and produce more pustules.

The fungus does great damage to the host plant by using water and nutrients needed by the developing kernels. As a result kernels are badly shriveled and wrinkled; many of them are so light they are blown out with the chaff in thrashing. Stems of infected plants are weakened and lodge or break over easily. Sometimes losses range up to 90%, and the crop is not worth harvesting.

Later in the season as the wheat plant approaches maturity, the red pustules turn black because the fungus forms black, winter spores (*teliospores*) instead of red spores. The teliospores remain dormant all winter on the straw and chaff. In the spring, with the onset of warm weather, the teliospores germinate to form *basidiospores* (*sporidia*) of two different mating types (+ and −). These tiny, colorless spores can infect only the leaves of native barberry. If a few basidospores from an infested field are blown onto a wet barberry leaf and germinate, the spores infect the leaf and form yellow pustules (*pycnia*), sometimes up to 10 mm in diameter, within a week or 10 days.

The *pycniospores* formed in pycnia of one mating type (+ or −), when splashed by rain or carried by insects, can fertilize the *receptive hyphae* in the pycnia of an opposite mating type. Cluster cups (*aecia*) form on the underside of the barberry leaf and produce myriads of yellow spores (*aeciospores*), which are carried by the wind to infect any wheat plants growing nearby. As many as 70 billion aeciospores may be produced on one large barberry bush. The aeciospores can infect only wheat. If these spores land on a wet wheat leaf, they germinate quickly and form a new crop of pustules containing red spores in 7–10 days.

Dispersal

The fabled winds of the Great Plains of North America, heralded in song and story, are both a blessing and a source of potential disaster to crops. Without the wind, there would be no harvests of wheat, oats, rye, or barley from the bountiful prairie lands. Wind is necessary for these crops because it carries tiny pollen grains from one flower to another, resulting in cross-pollination; fertilization follows within a few hours, and the developing seed enlarges rapidly as food is transported to and stored in the seed heads.

But the wind is also the principal agent for moving the invisible rust spores from one location to another. Local breezes spread red spores from one field to another to start new infections; of more importance are those spores wafted upward 3 km or more where they may be blown hundreds or thousands of kilometers before they descend with the falling rain.

If weather conditions are favorable in the spring, as warm weather advances, the rust multiplies, and the spores sweep northward on the prevailing winds. A heavy rust epidemic in Mexico and Texas threatens the wheat crop in Oklahoma and Kansas. These in turn supply the spores for infection in the Dakotas, Minnesota, the intermountain states, and Canada. Similar patterns of infection occur in Europe and Asia.

Spores blown southward in summer attack volunteer wheat, early-sown wheat, and certain grasses. The fungus overwinters as mycelium in such hosts and the next spring produces more red spores to begin another epidemic.

Management

Rust-Resistant Cultivars. The principal means of managing or containing wheat stem rust is the use of rust-resistant cultivars. Yet a resistant cultivar is useful for only about 5 years after its introduction before it becomes susceptible to the rust, and yields go down. Why? Well, there are several hundred parasitic strains or physiologic races of stem rust that differ in their ability to attack different cultivars of wheat. And more of these pathogenic races arise every year. As rapidly as scientists breed a new resistant cultivar, new fungus races arise that can attack it. There are three principal ways that new pathogenic races arise:

1. *Sexual matings*—When teliospores germinate, new combinations of genetic characters occur in the basidiospores. Upon fertilization in the pycnia, the different genetic characters are combined in aeciospores and result in the appearance of new races of the fungus. These new races may be able to infect new or old wheat cultivars that were previously immune to existing races.
2. *Hyphal fusions*—When red spores of different pathogenic races happen—by chance—to infect the same wheat leaf, some of the hyphae may fuse and exchange nuclei. As these hyphae grow, new pathogenic races result from the combined influence of the exchanged nuclei.
3. *Mutations*—Unexplained genetic changes occurring in the nuclei of a given race may change its parasitic capability, forming a new race.

Although new rust-resistant cultivars must be developed and introduced on a continuous basis, they are responsible in large part for the consistently high wheat yields obtained in the United States. Experts agree that without rust-resistant cultivars annual wheat yields in the United States would be much lower than they are. Moreover, resistant cultivars provide insurance against disastrous losses, which could seriously disrupt the milling industry, lead to possible shortages of wheat supplies, and result in higher prices for bread and other wheat products. Resistant cultivars, however, are not foolproof; other methods to help manage the stem-rust fungus must be sought.

Fungicides.   Before 1960 there was little or no use of fungicides to manage wheat rust. Fungicides were expensive; they did not give good enough management; and income per hectare from wheat was low. Therefore, the cost/benefit ratio for fungicides was not high enough to justify their use in managing stem rust on wheat.

The use of chemical sprays, however, became economically feasible with the introduction of ULV spraying and new fungicides, which permitted effective treatment of wheat fields with as little as 8 liters of fungicides per hectare applied by helicopter or airplane. Two applications of appropriate fungicide mixtures, coordinated with forecasts that predict weather favorable for rust epidemics, may reduce stem-rust damage as much as 75% compared with untreated fields. These chemicals both protect the plant and eradicate the fungus. Therefore, two sprays, the first when rust is just barely detectable and the second 10–14 days later, can give economic control. Systemic fungicides that are applied to the soil as granules just before sowing and again in mid-season as a spray are now available in some states.

Leaf Rust of Wheat

Leaf rust of wheat is found in every place that wheat is grown. Almost every year it causes damage in the southeastern United States and the Ohio and Mississippi valleys, where weather conditions are most favorable for its development.

Leaf rust usually attacks the leaf blades and sheaths—hence the name. Sometimes it occurs on the stem proper, just below the heads, and also on the chaff and beards. Leaf rust seldom shrivels the kernels, but it does reduce the size, number, and quality of grain. Furthermore, heavily rusted plants are weakened and dwarfed. In such cases, the yield may be reduced by as much as 90%.

There are many strains of the leaf-rust fungus (*Puccinia recondita* Rob. ex Desm.). The red spores are the only important ones. They are found all year on wheat in southeastern states. There the spores remain alive over the winter on volunteer wheat plants and fall-sown wheat. In the spring the spores are carried by wind to infect plants where the fungus may not have survived the winter. During warm moist weather a new generation of red spores is produced every week or 10 days.

To date, leaf-rust-resistant cultivars have been the principal means of reducing leaf-rust losses. However, for such cultivars to be of use, they must also be resistant to powdery mildew, another disease that is prevalent in the southeastern United States. Resistant cultivars must be replaced periodically because new fungus strains arise that can attack them, as described in the section on stem rust.

Cedar-Apple Rust

Apples and red cedars both grow well in the eastern United States. In the Midwest, where cedars do not grow naturally, the pioneers often planted them for windbreaks near orchards. Unfortunately, these early settlers did not know that both kinds of trees suffered from the same disease, cedar-apple rust. Where cedar trees are numerous and close to apple trees, rusts can cause serious losses of apples.

Symptoms

Rust diseases injure apples in several ways. Fruits of infected trees are deformed and ugly, and so reduced in grade that most of them are classified as culls. Considerable loss from secondary rots may follow rust infection. Fruit size is reduced by heavy foliage infection. Many infected leaves fall off by midsummer. The trees are weakened and may suffer winter injury. Young trees may die from infection and defoliation.

Disease Cycle

The fungus that causes cedar-apple rust (*Gymnosporangium juniperi-virginianum* Schw.) spends almost 2 years of its life cycle on cedar trees. Cedar leaves are infected in late summer or fall by aeciospores blown from lesions on apple leaves. Sometimes the aeciospores lie dormant in the leaf axils before infecting cedar trees the next spring. From these infections small brown galls, called cedar apples, appear during the summer but do not mature until the following spring, when they may grow up to 5 cm in diameter. After a few warm spring rains, about the time that apple blossoms are pink, the galls absorb water, increase in size, and squeeze out long, thin, bright orange, gelatinous tendrils that form a mass several times that of the original brown galls. These tendrils, or horns, produce two-celled teliospores that germinate by forming four sporidia (basidiospores). All teliospores do not germinate at the same time. With each rain the horns push out further and expose more spores. When the supply of spores is gone, the galls dry and may drop. Often they remain hanging on the cedars through the next year.

As the air becomes dry, mature sporidia are discharged forcibly into the air where they can be blown several kilometers. If they happen to fall on a wet apple leaf when the temperature is 15°–20°C, they germinate quickly and penetrate the leaf within an hour or so. Young developing apples also may be infected, usually near the blossom end.

Yellow rust spots (up to 10 mm in diameter) appear on the upper leaf surface within 1 to 3 weeks, depending on the temperature and the susceptibility of the apple. As many as 300 spots may occur on a single apple leaf. As the spots increase in size, a sticky gum containing pycniospores appears. Insects are attracted to this gum; they carry spores from one rust spot to another and thus fertilize the fungus. After fertilization the fungus grows through the leaf and produces fruiting bodies, called aecia, on the lower surface of the leaf. The aeciospores ripen in July and August and escape from the aecia. Errant winds blow the spores to nearby cedar trees to complete the 2-year life cycle of this rust fungus.

Management

In neighborhoods where apples are an important backyard crop, a simple, practical way to control cedar-apple rust is to cut down any cedar trees nearby. If apple orchards are in valleys fringed by cedar-bearing hills or if the esthetic value of cedars on some estates outweighs the value of nearby orchards, however, it may not be practical to cut down the cedars.

In the eastern United States it is difficult to grow apples profitably unless a rigid spray program for containing diseases and insects is used. Cedar-apple rust is one of the diseases that make this spraying necessary.

Apple cultivars differ in their susceptibility to cedar-apple rust. Whenever possible, resistant cultivars should be planted where cedar-apple rust is a problem.

Smuts

Until the twentieth century, smut diseases throughout the world were second only to rusts as the cause of serious grain losses in cereals. Today, total losses to smuts are low, but serious localized losses can occur, particularly in developing countries. Unlike the rusts, which attack the stems and leaves of cereal crops (and many other plants) smuts do damage by attacking the grain kernels. The reduction in yield brought about by these diseases is immediately observable since kernels of grain are replaced by the black, dusty spore masses of the smut fungi. Kernels that do develop may be covered with these dark, oily spores, and are, therefore, reduced in quality.

There are three basic types of smuts, which are distinguished by the way in which they infect a host plant. Corn-smut spores can penetrate the plant and cause disease at any growing point (*local infection*) including the tassel, leaves, and most importantly the developing corn kernels. Covered smut (also known as stinking smut or bunt) is caused by a fungus that infects young barley, oat, or wheat seedlings (*seedling infection*). Loose smut is caused by a fungus that infects the ovary of a cereal plant when the plant is in flower (*floral infection*).

Corn Smut

Symptoms

Minute galls form on leaves and stems of young seedlings infected with corn smut; infected seedlings remain stunted and may eventually die. When older plants are infected, localized areas such as the axillary buds, flowers of the ear (silks) or tassels, leaves or stalks will develop symptoms. However, most damage occurs when ears are infected (Fig. 13.2).

Fig. 13.2. Massive smut galls on corn ears. Sweet corn is especially susceptible to corn smut, which is caused by *Ustilago maydis* (DC) Cda.

The fungus penetrates growing tissue and grows throughout a section of the plant. Fungus invasion causes cells of the host to enlarge and to divide at an accelerated rate. The result is a swollen growth (*gall*), which is, at first, covered with a greenish-white layer of fungal tissue. As the gall becomes older and matures, the greenish-white membrane turns silvery white and the gall is filled with a mass of black, oily spores. When mature, a gall may range from 1 to 20 cm in diameter and contain millions of spores. When the gall ruptures this mass of black spores escapes into the air; spores can often be seen as black dust on plant parts near the ruptured gall.

When the galls are young and no hint of the black, sooty spores is found, the Indians of Mexico and Central America eat these young galls as a great delicacy. As with any fungus, however, one should always make sure precisely what the proper identity of the fungus is before it is eaten!

Causal Agent and Disease Cycle

Corn smut is caused by *Ustilago maydis* (DC) Cda., which produces two kinds of spores. Black spores (teliospores) are produced in the galls and overwinter in debris in the soil. Basidiospores are produced when the teliospores germinate in the spring. After the fusion of two basidiospores the resulting mycelium infects the corn plant, and symptoms develop.

Management

Some corn hybrids have a slight amount of resistance to corn smut. In home gardens, corn smut can be managed by simply removing galls when they are first observed and burying them. Crop rotation may also be helpful in small plantings of corn. Both of these management methods are of limited use in large corn-growing areas.

Covered Smut (Stinking Smut or Bunt)

Symptoms

Plants infected with covered smut may be a few to several centimeters shorter than healthy plants. Roots on infected plants are usually poorly developed. When the heads of infected plants emerge, the color of the head is usually bluish green instead of the normal yellowish green. At the time when the grain would normally be mature, the entire contents of the kernel is filled with a black, oily, dusty mass of spores. The fungus spores give off a distinctive odor of decaying fish.

Causal Agent and Disease Cycle

Covered smut is caused by three species of the genus *Tilletia,* which all have similar disease cycles. The fungi produce the same two types of spores as the corn-smut fungus. Teliospores are black, sooty spores that fill the grain kernels; when the kernels rupture, the teliospores fall to the ground or onto other healthy kernels. Spores that fall to the ground may survive for several years; spores that land on healthy grain stick to the outer surface of the seeds and overwinter there.

In the spring when infected seed of barley, oats, or wheat germinates, the teliospores also germinate to form basidiospores. After the fusion of two basidiospores, the resulting mycelium penetrates the seedling and grows between the plant cells. The mycelium reaches the growing point of the plant and continues to grow in pace with the plant. The mycelium grows up into the spike and into the developing kernels; here the mycelium penetrates the plant cells and is transformed into teliospores. The smutted kernel becomes filled with teliospores and the cycle is ready to repeat.

Management

Covered smut can be managed by using smut-free seed obtained from resistant cultivars or by using seed that has been treated with a fungicide. Care should be taken to avoid contamination of healthy kernels with the spores from smutted kernels. Also, since the teliospores can survive in the soil for several years and infect germinating seedlings, the use of a systemic fungicide is helpful. Extension service staff can provide the latest information on resistant cultivars and fungicides.

Loose Smut

Loose smut is caused by *Ustilago nuda* (Jens.) Rostr., which produces teliospores and basidiospores. The fungus overwinters as mycelium in the infected seed. When infected kernels germinate, the fungus also begins to grow between the plant cells until it reaches the growing point of the plant. The mycelium infects the spikelets and invades most of the cells in the spike except the rachis. As a result, when the head emerges the masses of teliospores are present instead of flowers.

Loose smut is difficult to observe until a plant has headed. One clue to the presence of loose smut is that infected plants grow faster and are taller than neighboring healthy plants, probably as a result of stimulation by the fungus. When the spike emerges, the masses of spores are covered by a thin, silver-white membrane, which quickly ruptures (Fig. 13.3). The spores are released

Fig. 13.3. Loose smut replaces healthy seed with black masses of fungal spores (*left and right*). Compare with healthy head (*center*).

at a time that coincides with the flowering of surrounding healthy plants. The wind carries spores to flowers of nearby plants. Here the teliospores germinate to form basidiospores. After fusion of two basidiospores, the fungus penetrates the flower and becomes established in the various tissues of the newly forming seed. The mycelium then becomes dormant and awaits the germination of the seed.

Management of loose smut is usually through the treatment of seed with a systemic fungicide. As with all smut diseases, the best management procedure is the use of disease-free seed.

Downy Mildews

There are many downy-mildew diseases that affect a wide variety of vegetables, ornamentals, cereals, other crops, and weeds, particularly in cool, humid environments. The diseases get their name from the downy growth, consisting of conidia and conidiophores, that cover the underside of infected leaves. Most downy-mildew fungi are obligate parasites. Most of them produce oospores within the host tissue. The mycelium invades all the leaf tissues. In asexual reproduction the mycelium produces branched conidiophores that emerge through the stomates of the host epidermis. The conidia are borne singly at the tips of the branches and, when mature, are readily dislodged.

The downy-mildew fungi are important plant pathogens. In some areas, it is almost impossible to grow a profitable crop unless they are adequately controlled.

Late Blight of Potato

One of the most tragic events in history was the Irish Famine in the middle of the nineteenth century. Two circumstances were responsible for this disaster. First, the impoverished population of Ireland, about four million people, had become almost wholly dependent on the potato for food. Second, the potato crops for 2 years in a row—1845 and 1846—were almost totally destroyed by late blight. The misery and suffering were extensive and terrible.

The Irish people never knew what hit them. They did not know that a microorganism—a fungus—had destroyed the crop. Some blamed it on lightning; others, on bad weather. Many believed God had put a curse on the nation. The English government was so alarmed that the Royal Horticultural Society offered a prize of 50 pounds to the person who first discovered the true cause of the disease. Rev. M. J. Berkeley won the prize when he published his beliefs that a fungus caused this terrible disease.

A million Irish people died from starvation or from disease following malnutrition in the wake of the late-blight epidemic. A million and a half more immigrated, many to America.

Late blight also contributed to the defeat of Germany in World War I. In 1917 it destroyed about a third of the potato crop, which made up a large part of the wartime diet of the Germans. Reduction of the already scanty food supply contributed to the breakdown in morale and physical endurance that forced the Germans to surrender.

Distribution and Importance

Late blight of potato occurs worldwide wherever potatoes are grown. Irish potatoes thrive in cool, moist climates, and these are exactly the conditions that favor growth of the fungus. In the northeastern United States and northwestern Europe, late blight is prevalent and destructive because cool, wet summers periodically occur in these areas and provide ideal conditions for late-blight epidemics.

Late blight kills the foliage and stems of potato plants and rots the tubers. Late blight may totally destroy a potato crop if no disease management measures are used. Usually though, losses vary and depend on the weather.

Late blight is also destructive on tomatoes.

Symptoms

An early symptom of late blight is brown, dead areas at the tips or margins of infected leaves (Fig. 13.4). If moist weather prevails, the entire leaf may be killed in less than 4 days. During dry weather the infection advances more slowly. Blighted leaves curl and shrivel when it is dry, whereas in wet weather they remain limp and soon decay with an offensive odor. Generally, the lower leaves are affected first; but in muggy weather when the entire plant can become black and wilted, the entire crop may be destroyed.

Often the late-blight fungus can be seen as a dirty white or gray mold on the underside of the leaves. The white growth consists of the hyphae and conidiophores of the fungus on which huge numbers of spores are produced. In dry weather the external fungus growth is seldom seen.

Fig. 13.4. Late-blight symptoms on potato leaflets. During cool, moist weather the entire leaf may be killed in less than 4 days.
Courtesy C. W. Averre

Late blight also affects potato tubers (Fig. 13.5). Not only does death of the plant tops reduce the size and number of the tubers, but tubers also may become infected from conidia washed into the soil or from hyphae in the mother plant. During weather favorable to the fungus, tubers may rot before harvest; during less favorable conditions, only a portion of a tuber may be infected. Diseased tubers may rot in storage or, if used as seed pieces, furnish inoculum for the succeeding crop.

Fig. 13.5. **Potato tubers sliced to show black tissue infected with the late-blight fungus. Such tubers rot in storage. If used for seed pieces they will start the disease in the next crop.**
Courtesy C. W. Averre

Causal Agent and Disease Cycle

Late blight of potato is caused by *Phytophthora infestans* (Mont.) d By. It has large thin-walled hyphae, without crosswalls, which grow in the intercellular spaces of the host. Club-shaped *haustoria* penetrate the cells and absorb food from them. Slender aerial branched conidiophores grow out through the stomates and produce lemon-shaped hyaline conidia (27–32 μm × 16–24 μm) (Fig. 13.6). The conidia germinate between 15° and 24°C. Below 15°C the conidia germinate by forming zoospores, and above 15°C by forming germ tubes. There are several pathogenic strains of the causal fungus.

The late-blight pathogen overwinters as mycelia in infected tubers. Volunteer sprouts that develop from infected tubers left in the field or thrown into cull piles become infected. The tubers sprout, and soon all aerial parts of the plant are invaded. In a few days conidia are produced. These conidia are splashed to leaves of adjoining plants or are blown to nearby fields where they germinate to start new centers of infection. Tubers near the ground surface are attacked during wet weather by emerging zoospores, which penetrate through wounds or lenticels. Harvested tubers also may become contaminated with conidia from infected leaves or soil. Occasionally tubers are infected by hyphae growing from the mother plant.

Late-blight development depends on the weather. The fungus sporulates abundantly at or near 100% RH between 16° and 22°C. Sporangia die in 3–6 hr below 80% RH. The conidia germinate in 30 min to an hour in free water at 10°–15°C, and penetration of host tissue may occur within 2 hr.

Temperatures above 30°C check growth of the fungus but do not kill it. When the temperature goes down, the fungus will begin to grow again.

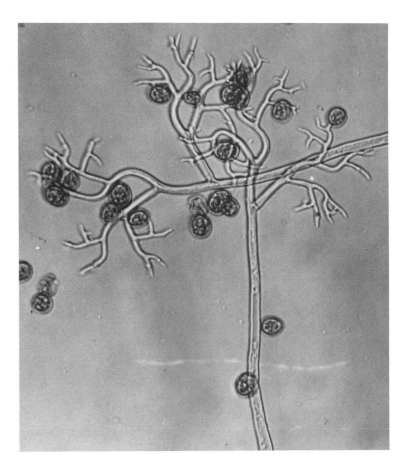

Fig. 13.6. Conidia and conidiophores of a downy-mildew fungus magnified about 530 times.
Courtesy USDA

Management

Late blight is such a destructive disease that potato growers must use a combination of practices if the disease is to be successfully managed. These include the use of sanitary measures, resistant cultivars, and effective fungicides. Moreover, late-blight management practices must be coordinated with an overall IPM program in order to produce the highest yield of a quality crop.

Growers should use only certified, disease-free potatoes for seed. All dump or cull piles should be burned or sprayed with herbicides before planting time. Likewise, volunteer potato plants should be destroyed in or near fields where a crop is to be planted. Use of the most resistant cultivars recommended by the state experiment station for a given area is important in curtailing late blight.

Generally, spraying with fungicides is necessary to manage late blight. Sprays should be applied, starting when plants are 15–20 cm high, every 4 to 5 days when the weather is damp or rainy. Proper timing and thorough coverage of foliage is necessary. Some states now have a late-blight warning or forecasting service that predicts when late blight will appear based upon rainfall, RH, temperature and sunshine data. Farmers should use such forecasting services to help them determine when to apply fungicides or other pest control measures.

Downy Mildew of Grape

Downy mildew of grape will go down in history as one of the first diseases to be controlled by a fungicide. And the discovery was accidental. In 1880, Dr. Pierre Millardet, a French plant pathologist, was studying downy mildew, which was destroying the grape vineyards in southern France. Millardet was a keen observer. One day as he walked in a vineyard in Bordeaux, he noticed that many vines along the roadside had not shed their leaves. They looked green and healthy. On the other hand, vines away from the road had lost all their leaves. Millardet asked the vineyard owner what had happened. The owner said that passers-by would pick the grapes alongside the road. To stop this pilfering, the owner placed some lime, bluestone (copper sulfate), and water in a bucket, stirred the mixture well, then took a broom and spattered the ugly blue-white liquid on the vines alongside the road. The farmer hoped the vines would look so poisonous and repulsive that thieves would not eat them. Sure enough, this psychological warfare was successful. The grape stealing stopped, and oddly enough the vines remained healthy.

Not only was Millardet observant, but he was also a man of action. He started mixing various proportions of lime and copper salts till he found an efficient combination that would protect grapes against mildew. It took 3 years of testing before he hit on the right combination for Bordeaux mixture, as it became known. For the next 50 years or more this remarkable fungicide was used all over the world to control not only mildew but also several fungal diseases on other crops, including late blight of potato. It would be difficult to calculate the worth of Bordeaux mixture. For decades it was the standard fungicide on potatoes, and it is still used to some extent.

Symptoms

Downy mildew affects all parts of the grape plant. It causes losses by killing leaf tissues and defoliation; it rots the ripening grapes and weakens, dwarfs, and kills the young shoots. The first symptoms are pale yellow, irregular spots on the upper leaf surface. On the underside of the leaf surface the downy growth of the fungus appears. The infected leaf areas are killed and turn brown. Often the leaves die and fall off. During favorable weather entire clusters of grapes may be rotted. The fungus thrives best during cool, moist weather.

Disease Cycle

Downy mildew of grape is caused by *Plasmopara viticola* (Berk. & Curt.) Berk & de T. Oospores (30 μm in diameter) overwinter embedded in dead leaves, which disintegrate and release the dormant oospores in the spring. Only oospores that have been frozen can germinate. They form sporangia on short stalks. Each sporangium (25 μm × 40 μm) produces about 20 zoospores (12 μm in diameter); these are blown or splashed to nearby leaves, which become infected within 1–2 hr if conditions are favorable.

Penetration usually occurs through the stomates on the underside of the leaves. The hyphae quickly spread through the spaces between the leaf cells.

The fungus obtains its food from the numerous globose haustoria that grow into the leaf cells and rob them of their contents. The cells begin to lose their green color as the chlorophyll is destroyed, the cell contents disintegrate, the cell walls collapse, and the entire cell turns brown.

Soon the mycelium begins to form sporangiophores, which emerge from the stomates. Great numbers of spores are produced on the treelike branches of the sporangiophores. They are blown or splashed in all directions. If they land on a wet grape leaf or shoot, they germinate quickly and form many zoospores. These zoospores then infect the plant through the stomates and lenticels, and the disease spreads rapidly. When conditions are favorable, a new crop of sporangia may be produced every 5–18 days, depending on the temperature, RH, and host susceptibility.

At the end of the growing season the fungus forms oospores inside leaves, and sometimes inside shoots and berries. Next spring the oospores start new infections.

Management

To grow grapes profitably, farmers must manage several diseases and insects. This IPM includes a season-long program of spraying or dusting. Downy-mildew control with proper fungicides should be fitted into the overall IPM program.

The most effective way to manage downy mildew of grape is to spray with fungicides at 7- to 10-day intervals beginning before bloom and continuing up to harvest. The time and number of applications depend upon the weather. The agricultural extension service in each state furnishes recommended spray schedules.

Proper pruning of the canes to permit free circulation of air and rapid drying of the leaves also helps manage downy mildew, as does plowing under or removing dead leaves. But the principal management measure is careful application of fungicides at the correct time in the correct way throughout the growing season.

Downy Mildew of Cucurbits

Downy mildew is a destructive disease of cucumber, muskmelon, and watermelon. Occasionally it also causes damage to gourds, pumpkin, and squash. All downy mildews thrive in moist weather, particularly when wet conditions prevail during the growing season. The downy-mildew fungus on cucurbits causes the most damage during warm, wet weather. It is particularly troublesome in the eastern and southern regions of the United States.

Symptoms

Infected leaves have yellow spots on the upper side; on the lower sides, the purplish mildew growth appears when the humidity is high. On cucumbers the affected leaves may die, and the whole plant may be severely stunted or killed. Maturing fruits fail to color properly, have no taste, and usually are sunburned.

Disease Cycle

No oospores of *Pseudoperonospora cubensis* (Berk. and Curt.) Rostow, which causes downy mildew of cucurbits have been found. Therefore, since the fungus is an obligate parasite, it must grow on cucurbits at all times of the year in frostfree areas. Conidia are blown northward in the spring and start new outbreaks wherever a susceptible host and favorable climate coincide. The conidia germinate by forming zoospores (10–13 μm in diameter). These penetrate through stomates and quickly form intercellular hyphae with globose haustoria that grow into the host cells and use the contents for food.

Different races of the fungus occur on the various cucurbit crops.

Management

Regular applications of fungicides will generally control downy mildew on cucurbits, although other measures also help manage the disease. Wild cucurbits growing in the vicinity where cucumbers or muskmelons are grown should be eradicated. Production fields should be isolated from one another, and resistant cultivars used, especially in small home-garden plantings where fungicides are not applied regularly. Harvested and abandoned fields should be disked promptly to avoid carry-over of inoculum. Rotations should be used. Late plantings should be isolated from areas where early crops were grown. Workers and equipment should stay out of fields when plants are wet from dew or rain. Resistant cultivars have been developed and should be used whenever possible. Lists of vegetables resistant to various diseases can be obtained from local extension service personnel.

Powdery Mildews

Plants with powdery mildew are so conspicuous and striking that almost everybody notices them. Powdery mildew is one of the oldest plant diseases known. Ancient writers regarded it with fear and superstition. It is mentioned in several books of the Old Testament; the ancient Greeks also wrote about it. Powdery-mildew fungi occur all over the world with the exception of the Artic and high mountain regions.

On some crops the disease occurs almost every year and causes great damage. These fungi attack all important crops including apples, cantaloupes, cereals, flowers, grapes, grasses, legumes, lettuce, muskmelon, ornamentals, peach, shade trees, tobacco, and many weeds. All broad-leafed plant species are host to some kind of powdery-mildew fungus. Only the pine trees and their relatives do not have powdery mildew.

Symptoms

Plants infected with powdery mildew have a striking appearance. Infected leaves and stems look as if they have been dusted with flour (Fig. 13.7). All aboveground portions of plants can be attacked, especially young leaves and shoots. Growing leaves become twisted and misshapen. Often infected buds

---

Fig. 13.7. Wheat leaves severely infected with powdery mildew. The black specks are perithecia, in which sexual spores are produced.

do not open; flower buds become discolored and dwarfed and finally die. On susceptible grasses and red clover, infected leaves turn yellow and brown with resulting unthrifty, stunted growth. Cattle do not like to graze on such forage. On cereal grains, powdery mildew may lead to severe yield losses, sometimes 20% or more, and the quality of the grain is reduced.

Powdery mildew can destroy an entire crop of grapes when temperature and RH are favorable for fungus growth. The grayish-white growth appears on the shoots, tendrils, cluster stems, petioles and upper surfaces of the leaves. In unusually favorable seasons, even the berries are attacked.

On roses the disease appears year after year. It reduces flower production and weakens the plants by attacking the buds, young leaves, and growing tips. Often infected buds do not open; flower buds become discolored, dwarfed, and ugly and finally die. Growing leaves become curled and distorted. On older leaves, large white patches of fungus growth appear.

Causal Agents and Disease Cycle

Three principal genera of fungi cause powdery mildew: *Erysiphe, Sphaerotheca,* and *Podosphaera.* Contrary to most fungi, which invade the host and grow inside, powdery-mildew fungi grow mostly on the outside surface of the plant. The white powdery coating on the leaves, twigs, and buds consists of white mycelium with slender, many-branched, septate hyphae. Many upright branches bearing chains of egg-shaped, colorless summer spores (conidia) arise from the network of hyphae (Fig. 13.8). At many points the hyphae are attached to the host surface by specialized branches called

Fig. 13.8. Erysiphe cichoracearum, a powdery-mildew fungus. A—Haustorium in epidermal cell. B—Conidiophore and conidia. C—Perithecia. D—Asci and ascospores. Magnified about 100 times.

Drawing by A. Husain after Smith and E. S. Salmon

haustoria. These rootlike organs grow into the epidermal cells and absorb liquids from the host sap, which the fungus uses for food. It is a very specialized and unique arrangement.

The powdery, fragile conidia are produced in great abundance. They are easily detached and blown by the wind; thus, extremely rapid spread of the disease can occur. The conidia germinate in less than 2 hr under favorable conditions, penetrate the leaves, and form haustoria. A new crop of conidia is produced every 5–10 days.

Perithecia are formed on leaves and twigs in late summer or as the crop matures. They are globose and have a hard, black outer coat. Within each perithecium may be found a small sac or spore case, the *ascus.* Within each ascus may be found two to eight *ascospores,* formed as a result of a sexual union. These sexual spores, safely protected inside the perithecial coat, act as overwintering spores. They can also give rise to new pathogenic strains.

The fungus can overwinter as mycelium in dormant buds. As soon as these buds begin to grow in warm weather, so does the fungus. It forms conidia, which are blown to nearby branches. Perithecia also endure through the winter. Warm temperatures and rain make the perithecia swell and burst. The winter spores (ascospores) are splashed to nearby leaves or buds where they germinate quickly to start new infections.

Environmental Effects

Both temperature and RH influence development of powdery mildew, but RH is more important. Most fungal spores need free water to germinate; however, free water is unnecessary for and will inhibit germination of powdery-mildew conidia. Maximum germination occurs at 97–99% RH and between 14°–24°C. Some germination occurs when RH is as low as 25–30%; however, conidia usually die quickly at very low humidities and at very high or very low temperatures.

Periods of low rainfall and moderate temperature favor development of powdery mildew, as do periods of high RH followed by dry conditions and/or warm days followed by cool nights. High nitrogen levels that produce succulent, rapidly growing tissue, easily attacked by the fungus, also increase mildew severity. Frequently, plants growing in shade also have more severe mildew. Reduced light apparently favors infection and fungus growth.

Management

Spraying and dusting with fungicides have been used for many years to manage powdery mildews. Under most conditions, weekly applications give good protection; during periods of frequent rains, active growth, or wide temperature changes, more frequent applications may be necessary.

Some fungicides when used frequently exert selective pressure on powdery-mildew fungi. As a result, strains resistant to the currently used fungicide appear, and poor management results. When this resistance develops, a different fungicide must be substituted for the one presently used. Some growers use two different fungicides to help prevent the occurrence of resistant strains. As one farmer said, "I use two or three different kinds of chemicals to keep the fungus from getting used to any of them."

Cultivars resistant to powdery mildew have been developed for some crops (apples, lettuce, small grains, and others). Growers should find out if resistant cultivars are available in their area and use them wherever possible.

Blast of Rice

Blast, one of the oldest known diseases of rice, has been recorded at least since 1637. At that time the trouble was thought to be due to heat absorbed into the grain during drying in hot sunshine. The heat or fever ascribed as the cause of the disease is perhaps the reason rice blast was called rice fever.

Rice blast is widely distributed and has been reported in some 70 countries throughout the world. Rice grown under irrigation in areas of low RH usually suffers much less damage than crops grown in humid regions.

Blast is considered the principal disease of rice because of its wide distribution and destructiveness. Seedlings are often killed. Panicle infection and leaf infection also cause serious losses. Current annual losses worldwide probably average about 3%; however, in some areas during epidemic years losses may average 50%.

Strains of the fungus that causes blast are also pathogenic on numerous other grasses.

Symptoms

The rice-blast fungus produces spots on leaves, nodes, and different parts of the panicles and grains, but seldom on the leaf sheath. The typical elliptical leaf spots have pointed ends. The centers are grey and the margin brown to red-brown. The spots usually begin as small water-soaked white to blue dots. They enlarge quickly under moist conditions on susceptible cultivars and may reach 1.5 cm long by 0.5 cm wide.

Leaf spots caused by the blast fungus often resemble and may be confused with those caused by other pathogens. Severely infected leaves become dry and brown. One of the most conspicuous symptoms is "rotten neck." These are lesions on the neck of the culm (stalk) and on the panicle branches near the panicle base. Rotten neck prevents the kernels from filling; often the weakened branches break over where the lesions occur. On highly resistant cultivars only minute brown specks are seen. The fungus secretes at least two toxic substances—picolinic acid and piricularin—that kill host tissues.

Causal Agent and Disease Cycle

Rice blast is caused by *Pyricularia oryzae* Cav. The grey, slender conidiophores are rarely branched (Fig. 13.9). They are produced on hyphae growing on or in the host tissue. Frequently the conidiophores emerge in clusters through the stomates of infected tissue. The grey conidia are variable in size and shape, but are usually club-shaped with two septae. From one to 20 conidia may be produced on a single conidiophore. Conidial size varies with the fungus isolate and environmental conditions.

The optimum temperature for mycelial growth, conidial production, and germination is about 28°C. Alternate light and dark periods are necessary for conidial production. The conidia germinate freely in water.

Fig. 13.9. **Conidia and conidiophores of *Pyricularia oryzae*, which causes rice blast. Magnified about 500 times.**
Adapted from D. E. Ellis

In temperate regions, mycelia and conidia on straw and seed are the principal overwintering agents, but the fungus also survives on winter cereals and weed hosts. Under dry conditions conidia will survive for more than a year, but under moist conditions they do not live long enough to infect the next crop.

Wind blows blast conidia for kilometers. On wet leaves they germinate quickly and penetrate the host within a few hours. The mycelium grows rapidly in host tissue and under ideal conditions produces a new crop of conidia in about 6 days. A typical leaf lesion may produce up to 6000 conidia a day for 2 weeks. Windblown conidia start new infections.

Environmental Effects

When rice seedlings are grown at 20°C, blast infection is more severe than at 28°C. Seedlings grown several days at 18°–20°C before inoculation are also more susceptible than those kept at 28°. Adult plants are more resistant at 28°C than at 18°C. In general, rice cultivars are more susceptible in dry soils than in wet soils. High RH favors disease development. In areas where there are frequent and long periods of rain, blast will cause severe damage. High nitrogen levels always increase disease incidence. Many workers have shown that rice plants with a high silica content, especially in epidermal cells, show slight damage from blast. Evidently, epidermal cells with high silica absorb less nitrogen and are more resistant to physical penetration by the fungus.

Management

Resistant cultivars have been used for many years as one of the principal methods of managing rice blast. There are many different pathogenic races of the fungus, and new races are continually occurring. Therefore, any given resistant cultivar is effective for only a limited period of time. New cultivars must be bred as old cultivars lose their resistance. Rice is such an important crop, however, that rice scientists continually monitor the prevalence of new races of the pathogen. So far, in many countries rice blast has been held in check by the continual breeding of new cultivars as the old ones break down.

In Japan where the rice crop is the principal food and is vital to the nation's welfare, fungicides have given practical management due to a combination of reasons. First, rice is a high-yielding crop, and the government guarantees a high price for the product. The Japanese also have an elaborate disease-forecasting service and a well-developed chemical industry capable of producing relatively cheap fungicides. Finally, the farmers in each neighborhood are well organized to carry out community projects. All of these factors combined permit timely, large-scale application of economical fungicides by aircraft even though individual farm holdings are small. Usually the cost of chemical management is less than 5% of the crop value, and the cost/benefit ratio makes chemical management economically feasible.

In addition to the use of resistant cultivars and fungicides, certain cultural practices also help manage rice blast. Planting can be timed to avoid periods ideal for blast development. Proper use of fertilizers, proper irrigation, and proper spacing of seedlings also help to decrease disease severity.

Gray Mold

Gray mold is one of the most common disease of field crops, fruits, ornamentals, and vegetables throughout the world. It can be devastating on greenhouse crops. Host plants include African violet, amaryllis, apple, artichoke, azalea, banana, bean, begonia, beet, cabbage, carrot, chrysanthemum, cucumber, cyclamen, dahlia, geranium, gladiolus, hyacinth, lettuce, lily, onion, peony, pepper, rose, snapdragon, squash, strawberry, tobacco, tomato, tulip, and others. Losses from gray mold are difficult to estimate but are in the hundreds of millions of dollars annually.

Symptoms

The disease derives its name from the obvious *gray-mold* layer of fruiting bodies produced by the causal fungus on infected tissues during humid conditions. The fungus can attack many plant tissues. It can blight flower petals (Fig. 13.10) and form a cobwebby growth of mycelium that can be seen with the unaided eye. If blossoms are infected, the unsightly blighted tissue may cling to the plant. Flower infections can eventually lead to fruit rots. After fruit or stem tissues are infected, they become soft and watery and appear light brown. In addition to petal blights and fruit rots, this fungus can cause leaf spots on certain hosts such as gladiolus, onion, and tulip. Sometimes the fungus also causes seedling damping-off. Disease symptoms may be present and most severe on plants or plant products in storage or, as with strawberries, on the grocery store shelf.

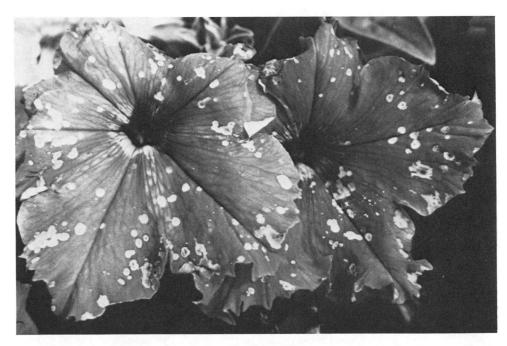

Fig. 13.10. **Petal blight of petunia caused by** *Botrytis* **spp. Each white spot probably resulted from infection by airborne conidia when the flowers were wet.**
Courtesy R. K. Jones

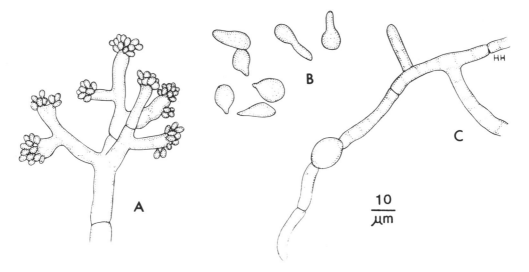

Fig. 13.11. The gray-mold fungi, *Botrytis* spp. A—Conidiophore with conidia. B—Conidia. C—Hyphal formation.
Drawing by H. Hirschmann after F. A. Wolf

Causal Agent and Disease Cycle

Several species of *Botrytis* cause gray mold; these are widespread as saprophytes, and many pathogenic strains are known. The fungus lives through dry conditions as black "rat-dung" sclerotia, which often develop in old decayed tissue. During wet weather the sclerotia germinate to form a gray mycelium which quickly produces fuzzy, gray masses of conidiophores bearing grapelike culsters of colorless, egg-shaped conidia averaging 10 μm × 5 μm (Fig. 13.11). These masses develop on the surface of rotted tissue during wet, cool weather or when RH is high. Conidiophores are often seen emerging from cracks in the epidermis of infected plant parts. The slightest air movement blows conidia to nearby plants to start new infections on wet, tender, weakened, injured or dying tissues. The fungus manufactures enzymes that soften or "tenderize" plant cells before they are actually invaded by the mycelium. This advance attack permits the fungus to grow more rapidly and increases the amount and spread of decay.

Management

Management of gray mold is difficult, but can be aided by the removal of infested debris from greenhouses, storage areas and fields. Cultural and storage conditions should allow for good aeration and the quick drying of surfaces of plants and stored plant products. Greenhouse RH should be reduced when possible by ventilation and heating. Repeated application of fungicides gives limited management of gray-mold diseases in the field and rots of fruits and vegetables in storage. Winter application of fungicidal dusts or drenches to soil underneath azalea or camellia plants, severly infected the preceding spring, will help kill overwintering sclerotia. Local extension service personnel can provide information on specific fungicide recommendations.

Cercospora Leaf Spots

Cercospora leaf spots are found on a wide variety of crops, particularly in warm, humid regions. On crops such as banana, beet (both red and sugar), celery, forage legumes, peanut, and soybean, yields can be much reduced by the disease.

Cercospora leaf spot of peanuts is one of the most important diseases of this crop world-wide, with annual yield losses of 15–50%. In the United States where chemical control measures are generally used, losses may be as high as 15%. The disease reduces the yield of nuts and lowers the quality of peanut hay. In addition, fallen leaves from infected plants provide organic matter to serve as a food source for other fungi, particularly *Sclerotium rolfsii*. As a result there is build up of inoculum, and the severity of stem rot is increased.

Many weeds and ornamentals act as hosts for this ever-present fungus.

Symptoms

Cercospora leaf spot of peanut is easily recognized by the brown or black circular spots on the leaves (Fig. 13.12). Often a bright yellow halo surrounds each spot. As the disease progresses, the spots enlarge until the entire leaf is affected. So many of these leaves fall off that often the ground beneath the plant is covered.

On the leaves of other crops (beet, celery, soybean, tobacco), the causal agent typically causes numerous small (2–15 mm in diameter) circular brown, tan, or dingy grey spots. Scattered through the centers of the lesions are little black dots composed of masses of conidia borne on short con-

Fig. 13.12. *Cercospora* leaf-spot of peanuts. Note the halo on the leaf lesions to the right.

idiophores. The spots may resemble a "frogeye," and *Cercospora* diseases are often called by this unusual name. Although *Cercospora* spp. primarily cause a foliage disease, infection can occur on any aboveground part of the crop.

During humid weather the spots become so numerous that they merge to form large irregular spots. At this degree of damage the leaves wither, die prematurely, and fall.

Under certain conditions it is difficult to identify the causal agent of a leaf spot simply by looking at it. During hot, dry weather frogeye lesions may be only pinpoint in size and difficult to recognize. On the other hand, during warm, humid weather the spots may enlarge rapidly. Frogeye lesions may be confused with spots caused by other fungi or bacteria. In addition, more than one kind of spot often occurs on the same leaf.

Causal Agent and Disease Cycle

Conidia of *Cercospora* spp. vary greatly in size. They range from 25–300 μm × 3–7 μm, with an average of 50 μm × 4 μm. Both temperature and food source affect conidial length. The slender, colorless conidia are straight to curved and have several crosswalls. A typical conidium is shaped like a baseball bat. The conidia are borne on brown, septate, knobby, sparingly branched conidiophores borne in clusters. The fungus sporulates and grows rapidly at 27°C.

There are many pathogenic strains of *Cercospora*. Some strains can attack more than one host. The fungus is so variable that it is often difficult to identify precisely.

These fungi survive from one season to the next on diseased leaves, stems, and seeds. When contaminated seeds are planted, they may produce weak, stunted seedlings with lesions on the cotyledons; in a few days the growing fungus begins producing conidia to start secondary infections (Fig. 13.13).

Fig. 13.13. Cercospora spp., the frogeye fungus. *Top:* Cross section of upper surface of leaf showing the fungus threads bursting through the leaf and producing spores on their tips. *Bottom:* germinating conidia entering the leaf through the stomata. Magnified about 300 times.
Drawing by H. Hirschmann after W. C. Sturgis

The conidia are blown or splashed by rain or carried by insects and machinery to nearby plants. They germinate within a few hours and quickly penetrate leaf tissue. If the weather remains warm and humid, new leaves become infected successively as they develop. Often it is difficult to find a leaf that has no lesions.

Cercospora leaf spot is more severe when susceptible crops are grown repeatedly in the same field without crop rotation. Early infection starts from conidia produced on the trash from last year's crop. Once the disease gets off to an early start, it becomes difficult to control.

Management

Use of good quality, disease-free seed helps in managing Cercospora leaf spots. Seed treatment with a fungicide aids in preventing seedling loss, but does not prevent development of the disease in the field. For transplant crops, such as celery and tobacco, fungicides should be applied regularly in the seedbed as part of an IPM program. To delay or prevent build-up of fungicide-resistant races of the fungus, it is a good practice to alternate or rotate the use of different fungicides.

The fungus is carried over from one season to the next on debris from the previous crop; therefore all trash should be plowed under (or burned) as soon as possible. Crop rotation should also be used to prevent inoculum build-up. On peanuts successful control of Cercospora leaf spots at present depends on regular applications of fungicides in the field.

Whenever possible, disease resistant cultivars (e.g., of soybeans and beets) should also be used.

Helminthosporium Leaf Spots

The three most important food crops in the world are wheat, rice, and corn. Species of *Helminthosporium* cause a number of serious leaf spots, seedling blights, root rots, culm rots, and seed infection on all three of these crops. Other crops including barley, forage grasses, oats, sorghum, and turfgrasses also suffer losses from these versatile fungi.

Southern Corn Leaf Blight (SCLB)

In the United States both the land area planted to corn and its total yield of grain are greater than for any other cultivated crop. About 80% of the corn crop is fed to livestock; the remainder is used in industry for human food. Essentially all diseases of corn have been or potentially can be managed by genetic resistance.

The history of SCLB in the United States presents a dramatic illustration of the dangers of genetic uniformity to modern agriculture. (See Chapter 6, Disease-Resistant Cultivars section, for additional discussion of this phenomenon.) The story starts in the 1950s with the discovery and introduction of male sterility to produce high-yielding corn hybrids at low cost. The source of male sterility, designated as Texas cytoplasm male sterility (T_{cms}), was transferred into many inbred lines of corn by backcrossing. Inbred lines containing T_{cms} were then crossed with lines possessing a genetic factor (a restorer gene) that restored pollen for kernel development. By 1970, the use of inbred lines and hybrids containing T_{cms}, was so widespread that about 85% of the hybrid seed corn sold in the United States was of this type.

In 1969, a disease of corn leaves and ears was observed in localized areas of Iowa, Illinois, and Indiana. The T_{cms} plants were severely diseased, whereas plants having normal cytoplasm showed no ear infection and only a few leaf lesions. Greenhouse pathogenicity tests demonstrated that the pathogen was a distinctive race of the fungus *Helminthosporium maydis* Nisik, which causes southern corn leaf blight. The old common strain of this fungus on corn was designated race O and the new strain race T.

In January 1970, the disease was reported as causing severe damage on T_{cms} corn in Florida. The spring of 1970 was unusually wet in much of the southern United States, and as a result the disease progressed rapidly northward from the South; by mid-July it was well established in the Corn Belt and a severe epidemic was occurring. In the South many fields of corn were a total loss. This epidemic of southern corn leaf blight (SCLB) received more publicity than any previous plant disease outbreak in the United States. Losses due to the epidemic were officially estimated at near one billion dollars in the nation as a whole. Average yield losses in the Corn Belt were 20–30%; southern states sustained much greater losses.

In 1971 the demand for the disease-resistant, normal-cytoplasm seed was far greater than supply. Race T of *H. maydis* survived the winter of 1970–1971 in most of the South, but weather conditions were dry and unfavorable to the fungus in the early spring and summer of 1971. Fortunately, the unfavorable conditions and the reduced usage of T_{cms} corn hybrids halted the SCLB epidemic, and a near record corn crop was produced that year. Sorghum, a major crop closely related to corn, is now in almost exactly the same vulnerable condition of genetic uniformity as corn was during the winter of 1969–1970.

Symptoms

Lesions on *Helminthosporium*-infected leaves are tan, spindle-shaped or elliptical, with yellow-green or chlorotic halos (Fig. 13.14). Later lesions often develop dark, reddish-brown borders and may occur on leaves, stalks, and husks. A black, felty mold may cover kernels in infected ears. Race O usually attacks only leaves.

Causal Agent and Disease Cycle

The mycelium of *H. maydis,* which causes southern corn leaf blight, is dark grey. The light green to brown conidiophores (60–280 μm × 6–10 μm), which arise in groups of two or three on the dead leaf spots, produce cigar-shaped, smokey grey conidia (40–150 μm × 11–27 μm); these have on the average six to eight crosswalls (Fig. 13.15). The spores germinate by sending

Fig. 13.14. Southern corn leaf blight, race T infection. *Left to right:* Sweet corn leaf with normal cytoplasm, popcorn and field corn leaves from plants with susceptible T_{cms}.
Courtesy Am. Phytopath. Soc.

out one germ tube each from the cells at either end of the spore, i.e., germination is bipolar. Numerous conidia are produced on the leaf lesions. The perithecia, or sexual stage, containing ascospores appear as tiny specks on or in disintegrating leaf tissue, but are rarely seen in nature.

The fungus overwinters as mycelium and spores in maize debris in the field and on kernels in cribs, bins, and elevators. Primary infections result from conidia carried by wind or splashing water to growing plants. Sporulation on lesions produces additional primary and secondary inoculum. Under ideal conditions the disease cycles in about 60–70 hr. The disease is most severe and spreads rapidly in warm (20°–32°C), damp regions. Race T is virulent on T_{cms} corn but causes only small lesions on corn having normal cytoplasm. This race produces a toxin that is host specific, i.e., it affects only T_{cms} corn. Race O does not differentiate between T_{cms} corn and normal cytoplasm corn.

Fig. 13.15. *Helminthosporium maydis,* which causes southern corn leaf blight. 1—Conidiophores. 2—Mature conidia. 3—Formation of conidia. 4—Conidium germinating. Magnified about 380 times.
Courtesy E. S. Luttrell

Management

Management of SCLB is currently achieved by use of normal cytoplasm corn hybrids and hybrids having disease-resistant sources of male sterility. Plowing under crop residues can reduce early infection by limiting survival of the fungus. Fungicides have been shown to reduce disease spread and yield losses, but genetic control with resistant cultivars offers the most desirable and least expensive management strategy. The use of certified seed that has been treated with a fungicide is always a good practice.

Brown Spot of Rice

Brown spot of rice is also called sesame leaf blight because the lesions are about the size of sesame seeds. The disease, which has been reported in all rice-growing countries, causes a seedling blight, weakens older plants, and lowers grain quality and weight. Brown spot was considered to be a major factor contributing to the Bengal (India) famine in 1942.

Brown spot is associated only with soil deficient in certain nutrients. The disease serves as an index of such conditions.

Symptoms

The most conspicuous symptoms of rice brown spot are the spots that appear on leaves and glumes; they may also appear on the coleoptile, leaf sheaths, panicle branches, and stems, but rarely appear on roots of infected plants. Typical spots on the leaves are brown, oval, and evenly distributed over the leaf. On susceptible cultivars the spots may reach 1 cm in length. Sometimes the spots are so numerous that the leaves dry out before the plant is mature, and as a result yields are drastically reduced. Under warm, humid conditions dark brown conidiophores and conidia make the spots look like velvet.

Causal Agent and Disease Cycle

Brown spot of rice is caused by *Helminthosporium oryzae* B. de Haan. The dark brown, abundant, many-branched hyphae grow profusely on diseased tissue. Grey conidiophores arise from the hyphae and give rise to slightly curved, cigar-shaped grey to brown conidia (35–170 × 11–17 μm) with an average of eight cells. Mature conidia germinate with two polar germ tubes. Many strains exist that differ in morphology, physiology, sporulation, and pathogenicity.

The brown-spot fungus overwinters in plant trash and infected grain. The conidia can survive 2–4 years in infected seed. Primary infection through diseased seed is common, but infection from trash and soil occurs just as frequently. Conidia germinate on wet leaves to form *appressoria* and infection pegs that penetrate the leaf cells. The pathogen produces a toxin that kills the host cells.

Environmental Effects

Brown spot is more severe on soils deficient in nutritional elements or on those in which toxic substances have accumulated. Deficiencies in potassium, manganese, magnesium, and especially nitrogen late in the growing season all increase brown-spot damage. Plants in soils with high levels of hydrogen sulfide also are prone to brown spot. Often brown spot and deficiency symptoms are difficult to distinguish in the field.

The optimum temperature for infection is about 25°C at RH above 90%. Mycelial growth is greatest at 27°–30°C, although the fungus grows over a wide temperature range.

Management

In areas where infection of seed with brown spot is common, seed treatment with fungicides reduces seedling damage. Where possible, rice should be grown only on soils with a proper nutritional balance; those with mineral deficiencies should be avoided. Proper fertilization is important for management of the disease. Field sanitation, crop rotation, adjustment of planting dates, and good water management help keep brown-spot losses to a minimum. Resistant cultivars should be used where needed.

Alternaria Leaf Spots

Fungi belonging to the genus *Alternaria* are commonly found on leaf spots and dead tissue. They are easily isolated from plant trash and seeds. Their identification is not easy because of the large number of strains and their ability to attack different hosts. *Alternaria* spp. cause leaf spots of apple, carnation, carrot, crucifer, cucumber, sweet pepper, potato, pumpkin, radish, tobacco, and tomato. Many weeds also suffer from leaf spots caused by *Alternaria* spp. Not only does the fungus attack crops in the field, but it also causes blemishes and transit and storage rots on cabbages, carrots, onion bulbs, peppers, potato tubers, and tomatoes.

On potatoes and tomatoes, Alternaria leaf spot is often called *early blight* to distinguish it from *late blight* caused by *Phytophthora infestans*. This name is not always accurate, as early blight often appears late in the season.

Symptoms

Leaf lesions caused by *Alternaria* spp. first appear as small watersoaked spots that quickly enlarge to form circular lesions of dead tissue up to 1 cm or more in diameter (Fig. 13.16). In humid weather, the black conidia on the surface of lesions become visible. Often they are arranged in concentric rings to give a "target spot" effect. Linear brown to black spots appear on petioles and stems. As is the case with other leaf-spot diseases, Alternaria leaf spot often increases in severity as the season progresses and leaves become mature. Often leaves die and drop off prematurely, and yields are decreased as a result of the plant's decreased ability to carry on photosynthesis. Typically the lower, older leaves show the most pronounced symptoms.

Fig. 13.16. Early blight on Irish potato leaflet, caused by *Alternaria* spp. Note the circular lesions with concentric rings and the brown dead tissue.
Courtesy C. W. Averre

Fig. 13.17. Scanning electron micrograph of *Alternaria* spp. conidia on leaf surface, magnified approximately 730 times.
Courtesy G. Van Dyke and H. W. Spurr

During transit and storage, crops such as carrot, cabbage, and potato are also injured by *Alternaria* spp. The outer leaves of cabbage heads are often overrun by the sporulating fungus and turn black. A black mold, consisting chiefly of conidiophores and conidia, develops on the surface.

Causal Organism and Disease Cycle

The much-branched mycelium of *Alternaria* species has numerous crosswalls and becomes dark with age. The short conidiophores produce tennis-racket-shaped, brown to black conidia which vary greatly in length, averaging 30 μm × 12 μm (Fig. 13.17). The conidia may or may not have beaks (some call them tails). Conidial production depends upon the isolate, light, moisture, and temperature. Several toxins produced by *Alternaria* species have been identified, and their role in premature yellowing and death of host tissue has been determined.

Alternaria species are carried on seed and plant refuse. They also overwinter as mycelium in woody debris such as stems and stalks. Conidial production starts with warm spring weather, and the spores are scattered by wind and water. Infected transplants also spread the fungus. The spores germinate readily and grow over a wide temperature range (4°–35°C). During warm, wet weather with prolonged periods of high RH or heavy dews, numerous lesions appear on leaves, petioles, and stems. Conidia produced on these lesions build up to high numbers as the season progresses. After the host plant dies, the fungus persists as mycelium in dead host tissue.

Management

Since *Alternaria* species are often carried on the seed, seed production in dry areas helps to minimize seed infection; however, seed treatment also is often desirable, especially on crucifers. Hot-water treatment at 50°C for 30 min is satisfactory for treating seed of members of the cabbage tribe.

Transplant crops should be sprayed with a fungicide in the seedbed to prevent transportation of the fungus on diseased transplants. Periodic spraying or dusting with fungicides is the principal method of managing Alternaria leaf spot diseases. Often repeated applications are necessary; use of fungicides must be incorporated into an overall IPM program as recommended by the local extension service.

Crop debris should be destroyed or plowed under as soon as possible and rotations used where possible. Maintaining proper fertility will help prevent severe damage on some crops.

Anthracnose

The name *anthracnose,* derived from a Greek word meaning ulcer, apparently refers to the ulcer-like lesions caused on many hosts. Anthracnose fungi are worldwide in distribution. Their many hosts include alfalfa, bean, coffee, clovers, corn, cotton, cucurbits, grasses, lupines, onion, ornamentals, peach, small grains, soybean, tobacco, and watermelon.

Anthracnose is a common and widespread disease that infects the foliage and fruits of many crops, particularly in warm, humid regions. On alfalfa, clover, and lupine anthracnose is one of the most destructive diseases in warmer regions. It reduces yields of hay and seed and can destroy stands of clover and other pasture crops.

Anthracnose fungi thrive where wet, warm growing seasons prevail. It is almost impossible to produce anthracnose-free seed in such regions. Therefore certified seed must be produced in semiarid, irrigated areas such as the western United States where rainfall is very low during the growing season.

Symptoms

Anthracnose may occur on plants at any stage of development. The appearance of the disease varies somewhat on different hosts. It develops most commonly on the young, succulent parts of stems, petioles, and fruits, but is not limited to them.

On beans the disease appears on all aboveground parts of the plant, but rarely on the roots. Usually the lesions are dark brown; in moist weather pinkish spore masses can often be seen. Elongate spots appear on the veins on the lower side of the leaf and eventually spread to the top side. Petioles, stems, and hypocotyls are often affected, causing the plant to collapse or damp off. The most striking symptoms occur on the immature bean pod. Small brown spots appear, which enlarge rapidly up to a diameter of 1 cm. They become dark brown to black and are depressed in the center. Sometimes the lesion penetrates through the pod, and the developing bean becomes infected.

On alfalfa and clover the first noticeable symptoms are usually dark brown or black spots on the petioles. The spots soon cut off translocation to the upper parts of the leaf, causing the leaflets to wilt and die. Dark tufts of black setae, (hairlike structures) are often conspicuous in the older lesions and help confirm a diagnosis of anthracnose.

On cucumber leaves the spots commonly start on a vein and expand into brown spots reaching 1 cm or more in diameter. Growing leaves may become distorted, and the fusing of many spots may blight entire leaves. Petioles and stem lesions are shallow, elongate, and tan. Lesions appear on ripening fruits as roughly circular, sunken, watersoaked areas, which turn black (Fig. 13.18). In the center, pink spore masses interspersed with black setae can be seen with a hand lens.

Fig. 13.18 **Mature cucumbers ruined by anthracnose.**
Courtesy of S. F. Jenkins, Jr.

Causal Agent and Disease Cycle

Anthracnose diseases are caused by species of *Colletotrichum.* Typically they have septate mycelium, colorless when young, dark when old. Often the mycelium forms masses of thick-walled cells called *stromata.* From these the fruiting structures, called *acervuli,* arise. Typically the acervuli form in the host tissue just below the epidermal cells, causing them to rupture. Short, oblong, one-celled, colorless conidia (5 μm × 20 μm) form on short conidiophores, and among them black hairlike *setae* are produced. Often the spores are so numerous that they form pink, shiny masses.

Anthracnose fungi live over from one growing season to the next on infected plant debris and on seed. When infected seed germinate, spores produced on infected cotyledons serve as secondary inoculum sources to infect other parts of the growing plant.

Spattering and windblown rain spread the spores to nearby plants. If a conidium lands on a wet leaf or petiole, it germinates and penetrates the surface by a short hypha that forms an infection peg to force its way through the cell wall. Optimum conditions for infection occur at about 20°C when RH approaches 100%. There are many strains of *Colletotrichum* species differing in their pathogenicity; thus breeding of resistant cultivars is a continuing task.

Management

One of the important ways to manage anthracnose is to use anthracnose-free seed. If necessary, seed should be treated with a suitable fungicide. Often certified seed dealers sell seed that has already been treated.

Rotations of 2–3 years and field sanitation help eliminate the fungus on overwintering trash.

Row crops should be cultivated only when dry if the disease is present because cultivation when foliage is wet spreads spores from one plant to another. Fungicides used in the overall complete disease management program also are of help. Resistant cultivars should be used if they are available.

Apple Scab

Apple scab—the most important disease of apples—occurs wherever apples are grown, except in very dry or warm climates. In years when scab is severe, crops may be a total loss unless management measures are applied. Even the most efficient management programs may fail to give satisfactory results when the disease is severe or weather conditions hinder the application of fungicides.

The primary effect of scab is a reduction in fruit size and quality. But the disease also causes premature fruit drop, leaf fall, and poor fruit bud development for the next year's crop. Scab also increases the cost of production, for efficient scab management requires timely and repeated fungicide application. Scab is such an important disease that several states now have established scab-forecasting services to tell farmers when conditions are favorable for scab development. Fungicides can then be applied at the proper time for the most efficient management.

Symptoms

Scab spots usually appear as small, olive areas on the sepals or young leaves (Fig. 13.19) of flower buds. Soon the lesions become olive green and assume a velvety appearance caused by the abundant production of spores on the ends of short, erect, threadlike branches of the fungus. Later the circular lesions turn black; some may reach 1 cm or more in diameter. Often the lesions are so numerous that they coalesce.

On the fruits, black circular spots (scabs) with deep cracks frequently develop, causing malformations that make the fruit worthless for market. Late-season infections result in small lesions, which enlarge into dark scab spots during storage.

Fig. 13.19. **Numerous scab lesions on the underside of apple leaves.**
Courtesy C. N. Clayton

Causal Agent

The scab fungus, *Venturia inaequalis* (Cke.) Wint. emend. Aderh., has both asexual and sexual spores. Either kind can germinate in water on a susceptible part of the apple tree and penetrate the cuticle of the plant by a microscopic peglike growth. Hyphae then grow between the cuticle and the cellulose wall of the outer layer of cells of the apple plant until a solid layer of fungal growth, which may be several cells deep, is formed over the lesion surface.

The asexual spores (conidia) are produced on short, erect, brown conidiophores, which successively form several one- to two-celled, red-brown, ovate to lanceolate conidia, averaging 16 μm × 8 μm. The sexual spores (ascospores) are produced in the spring in fruiting bodies (perithecia) of the fungus that develop in dead apple leaves on the ground. The tiny, black perithecia (90–150 μm in diameter) are spherical and have a small opening (*ostiole*) at the top through which spores are discharged (Fig. 13.20). Inside each perithecium, 50–100 asci are formed. Each hyaline ascus is shaped like a bag and contains eight ascospores (13 μm × 6 μm), each consisting of two cells of unequal size. The ascospores are colorless at first but turn brown when mature.

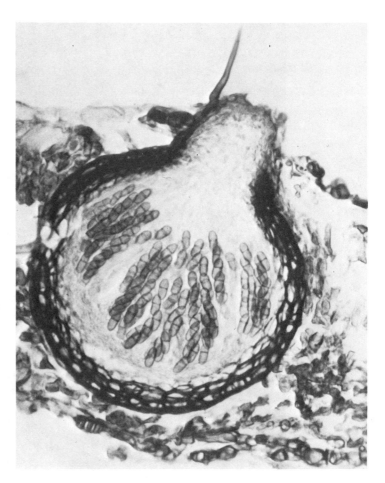

Fig. 13.20. Perithecium of the apple-scab fungus containing mature ascospores, magnified about 750 times. The ascospores mature in the early spring and are discharged into the air. If they fall on a wet apple leaf or flower, new infections begin.
Courtesy J. R. James

Disease Cycle

Perithecia of the scab fungus overwinter in dead leaves on the ground. Cool fall weather, snow cover, and a cool, moist spring favor the production and discharge of ascospores. In areas favorable for scab development, the ascospores are ordinarily mature and ready for discharge during rainy periods by the time the first susceptible apple buds start to open in the spring.

When dead leaves with perithecia become thoroughly soaked in the spring, the asci elongate, push through the ostiole, and forcibly discharge the ascospores into the air. The ascospores are carried by air currents to nearby leaves and flowers. For infection to occur at 5°C, the ascospores must be continuously wet for 18 hr; at 18°–24°C, they must be wet only for about 6 hr. Not all perithecia ripen at the same time, so ascospore discharge may occur continuously over a period of 4 weeks or so. This discharge often happens just at the time when apple leaves and sepals are expanding rapidly.

The ascospores penetrate host tissue by means of peglike hyphae much like the conidia; the developing mycelium absorbs nutrients from the leaf cells and produces enormous numbers of conidia within 8–15 days. Conidia are easily detached when wet with rain and are washed or blown to other leaves and fruits to establish secondary infections, especially during cool, wet weather.

After infected leaves fall to the ground, they die; and the hyphae form perithecia, which survive during the winter to start the disease cycle again the following spring.

Management

Although the breeding of scab-resistant apple cultivars holds promise for the future and some cultivars in use today exhibit some resistance (Fig. 13.21), the most important method of scab management is timely spraying with the proper fungicides. You cannot grow pretty, unblemished apples unless you spray. But fungicides are expensive and must fit in with other aspects of orchard management. Therefore, apple growers should consult their state agricultural extension service for the latest information on managing apple scab.

The spray programs most widely used depend chiefly on protection of the susceptible parts of the apple tree from infection by the fungus. As new leaves expand and fruits enlarge, they must be protected with fungicides. This protection should begin before the first infection occurs in the spring and should continue through repeated application of fungicides as long as there is danger of infection. Thus a scab management program may require 6 to 10 fungicide applications per season, beginning in the spring when weather conditions become favorable for infection and continuing almost until harvest. Scab-forecasting services, based on computer analysis of weather conditions, are becoming more accurate each year, allowing many farmers to apply fungicides at more precise intervals or when conditions ideal for scab outbreaks develop. Several fungicides give excellent scab management. All can act as protectants that prevent infections, and some have the ability to stop infections that have already started; others can eradicate young scab spots.

Fig. 13.21. A—The scab-resistant apple cultivar 'Liberty.' B—The scab-susceptible cultivar Delicious.
Courtesy of T. B. Sutton

Treating Fruit Trees at Home

Every home grower who has a few apple, peach, and pear trees dreams of producing fruits as beautiful as those pictured in the catalogs. Such is rarely accomplished. To produce umblemished fruit of highest quality, home gardeners would have to follow the same exacting spray schedules required of commercial growers who produce fruit for market. However, satisfactory results in home plantings can be obtained with a simpler spray program than that required in commercial orchards.

Repeated applications of an "all-purpose" spray mixture will give a fair degree of management of most diseases and insects that attack fruit trees. Ready-mixed commercial packages of effective pesticides can be purchased under various brand names, or the separate materials can be purchased and combined. *In either case, read the labels on containers to determine contents and directions for use.*

Apply the first spray as new buds open and continue at 2-week intervals until about 2 weeks before harvest. *Do not spray when trees are in bloom* except with streptomycin. Application of sprays containing streptomycin at 100 ppm (1 tsp/gal) at 5-day intervals during bloom will help control fireblight (see Chapter 16).

Missing applications or poor coverage will reduce the effectiveness of sprays. To boost the spraying effect: (1) rake up rotten fruits and dead leaves from under trees and burn or bury them; (2) cut off and burn dead and injured twigs and branches.

The chemicals recommended for treating fruit trees change from time to time as new chemicals are developed and become available. The homeowner should keep up with the latest information on what chemicals to use.

Rhizopus Soft Rot

A common bread mold is caused by species of *Rhizopus,* which also affect fleshy organs of flower crops, fruits, and vegetables, usually during storage, transit, and marketing. Rhizopus rot is the most destructive disease of sweet potatoes in storage (Fig. 13.22) and also causes losses on cherries, corn, cucurbits, gladioli, peaches, peanut, squash, and tulips. As a storage rot of tobacco, *Rhizopus* species must be guarded against every year.

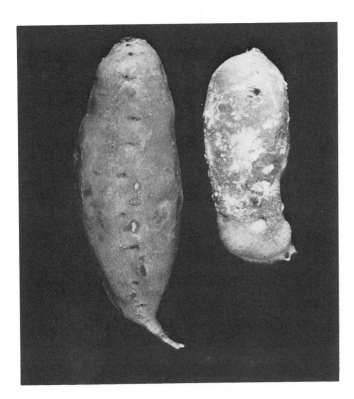

Fig. 13.22. Rhizopus rot on sweet potato (*right*) and disease-free sweet potato (*left*). Note the fuzzy mycelium of the fungus on the diseased sweet potato.
Courtesy of J. W. Moyer

Soft rot losses vary with the crop, storage temperature, RH, amount of wounding of the fleshy organs or fruits, and amount of inoculum present. When conditions are favorable, losses can be great in just a few days.

Symptoms

The first symptom of infection by *Rhizopus* species is a soft, watery rot, which progresses rapidly. A squash fruit may become completely rotted in 2–3 days; a sweet potato root, in 4–5 days. If the skin of infected tissues is unbroken, the softened fleshy organ gradually loses moisture and shrivels into a mummy. More frequently, the softened skin ruptures, fluid leaks out, and tufts of whisker-like, grey sporangiophores develop rapidly. The bushy growth extends to healthy parts of the affected fruit, to adjacent fruits, and sometimes to the wet surface of containers. At first the rotting fruits have a mildly pleasant smell, but soon a sour odor develops. Affected fruits sometimes dry out rapidly, and dry rots develop.

Causal Agent and Disease Cycle

Rhizopus spp. are found almost everywhere in nature and can use many kinds of crop debris as food. The mycelium has no crosswalls. When the conspicuous aerial mycelium grows on a surface, it produces arching aerial branches (*stolons*). Wherever they touch the surface, the stolons anchor themselves with rootlike hyphae (*rhizoids*). Opposite the rhizoids, erect hyphae grow in tufts up into the air. Each aerial hypha becomes a sporangiophore, which produces a black, spherical sporangium at the tip containing thousands of tiny black sporangiospores (5μm × 5 μm). When the sporangium breaks open, the sporangiospores float about in the air (Fig. 13.23). If they fall on a moist surface or wound of a susceptible plant organ, they germinate and produce mycelium again.

When unfavorable environmental conditions (e.g., low or high temperatures, diminished food supply) occur, the fungus produces black, warty sexual spores called *zygospores,* formed by the fusion of two hyphae. The zygospores are resistant to adverse environmental conditions and serve as the overwintering stage. Each zygospore germinates by forming a short hypha bearing a sporangium.

Both zygospores and sporangiophores may survive for many months. The fungus is such an efficient saprophyte that it undoubtedly subsists on crop debris between crop seasons. The optimum temperature for fungus growth and fruit decay is about 25°C with 75–85% RH.

Rhizopus spp. are strictly wound parasites. The hyphae produce enzymes that dissolve the cementing substances (*pectins*) between walls of plant cells. The cells become unglued from each other, and a "soft rot" develops. The fungus also produces other enzymes that break down the cellulose of cells, which then disintegrate. The fungus does not invade cells until after they have been killed by the enzymes; that is, the fungus kills cells first, then feeds on the dead remains like a saprophyte.

Soft rot develops quickest on mature fruit. Unripe fruits do not rot rapidly.

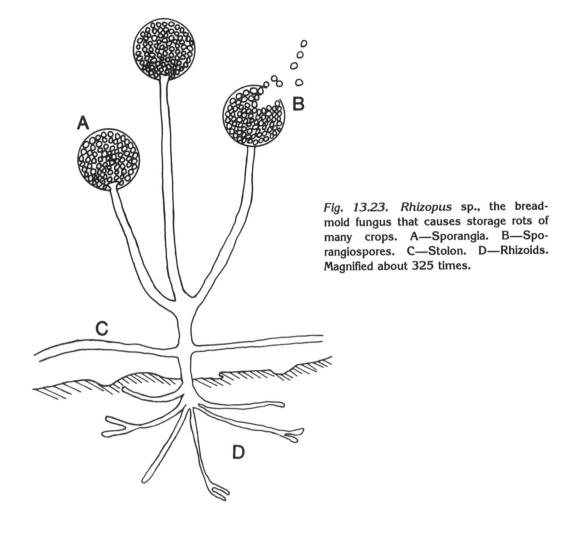

Fig. 13.23. *Rhizopus* sp., the bread-mold fungus that causes storage rots of many crops. A—Sporangia. B—Sporangiospores. C—Stolon. D—Rhizoids. Magnified about 325 times.

Infection and invasion of tissues are also slowed by cool temperatures (below 20°C) and low RH.

Management

One way to reduce Rhizopus soft rot is to avoid wounding fleshy fruits, roots, and tubers as much as possible during harvest, handling, and transportation. Wounded fruit should be discarded or packed and stored separately. Sanitation is important. Storage containers and warehouses should be cleaned of crop debris before use and disinfected.

Very succulent fruits, such as strawberries, should be stored at about 12°C. Sweet potatoes may be protected by holding them at 25°–30°C and 90% RH for 6–8 days; during this period the cut surfaces cork over so the fungus cannot penetrate. After the "curing" period, the storage temperature is lowered to 12°C.

Good management is also obtained by wrapping individual fruit in paper impregnated with fungicides.

14

Fungal Diseases of Shade and Forest Trees and Decay in Wood

The shade and forest trees of North America are a precious natural resource that we must protect and treasure. These silent, majestic inhabitants cover glens, glades, and mountain slopes, roof over cathedral groves in hidden valleys, anchor hillsides against the wash of sudden downpours, shelter the violets and wild columbine, purify the freshet as it rushes toward the lazy rivers, provide thickets filled with birds' song and hidden animals, supply lumber and paper, splash vistas with mosaics of orange, yellow, gold, and red in autumn, and etch bare trunk and limb against the deep mantle of winter snow. Trees restore the air for human use, and restore our souls with their beauty, diversity, and nobility.

Like cultivated crops, trees can be attacked by pathogenic fungi, which stunt their growth and may eventually kill them. Management of tree diseases is difficult, in part because the host plants are widely dispersed. Three major diseases—chestnut blight, Dutch elm disease, and oak wilt—have severely devastated three of our most important hardwood species during the past century. The fungi causing these diseases were introduced to the United States from other countries, where they were of relatively minor importance. However, in this country they spread rapidly, causing widespread destruction.

Softwood trees are not immune from fungal diseases. Infection of pine

Table 14.1. *Some Important Fungal Diseases of Shade and Forest Trees*

Disease	Pathogen	Host(s)
Chestnut blight	*Endothia parasitica*	American chestnut
Dutch elm disease	*Ceratocystis ulmi*	American elm
Oak wilt	*Ceratocystis fagacearum*	Many oaks
Annosus root and butt rot	*Heterobasidion annosum*	Coniferous trees
Fusiform rust	*Cronartium fusiforme*	Oak/loblolly, slash, longleaf and short-leaf pines
Little leaf	*Phytophthora cinnamomi*	Pine
Decay	Many fungal species	Wood in buildings

trees, in particular, can cause serious economic losses because of the importance of pine species in the pulp and lumber industries. Because pine trees are often grown in plantations (especially in the South), management of pine diseases is more feasible than it is in natural forests.

Even after trees have been cut and processed into lumber, the fungi do not give up. Wood decay induced by fungi may reduce the appearance and lifespan of wood structures. Fortunately, wood decay can be prevented or reduced by various measures.

A summary of the diseases discussed in this chapter is presented in Table 14.1.

Chestnut Blight

The American chestnut, a valuable hardwood species, has been virtually eliminated by chestnut blight throughout its natural range on the upper slopes of the Appalachian Mountains. In 1900, this handsome, useful tree (Fig. 14.1)

Fig. 14.1. Chestnut timber, Great Smokey Mountains, western North Carolina. In 1900 this majestic hardwood grew in almost pure stands for mile after mile throughout the upper slopes of the Appalachian Mountains of eastern United States. By 1940 the chestnut-blight holocaust, caused by an insidious fungus, had virtually eliminated this useful tree over thousands of square miles of its natural range.
Courtesy Forest History Society

grew in almost pure stands for miles on end. By 1940, nothing remained but the dried trunks and bare racks of a few dead monarchs high up on some inaccessible hillside.

The fungus *Endothia parasitica* (Murr.) P. J. and H. W. Anderson that causes chestnut blight evidently was introduced unknowingly into New York on nursery stock about 1904. It spread to the highly susceptible American chestnut and within 30 years thousands of square kilometers of chestnut trees were wiped out.

Symptoms

The first conspicuous evidence of chestnut blight is brown shriveled leaves on twigs that have been girdled by bark lesions. Swollen or sunken cankers forming on twigs, branches, and the main trunk quickly kill all tissue beyond the canker. Young cankers are yellow-brown in contrast to the normal olive-green of the bark. As the cankers become older, the bark splits and falls away. In older cankers, red-brown spore pustules up to 2 mm in diameter form.

Causal Agent and Disease Cycle

The chestnut-blight fungus produces summer spores in pycnidia; the colorless, one-celled oblong spores (averaging 8.6 μm × 1.6 μm) are squeezed out like toothpaste from a tube, following warm summer rains. These spores stick to the beaks and feet of bark-feeding birds. Bark beetles also become contaminated. Both birds and bark beetles can carry the spores considerable distances to infect other trees. Winter spores puffed in tiny clouds from other bark cavities are blown to nearby trees. Both kinds of spores infect through feeding wounds made by birds and bark beetles. Hyphae from germinating spores penetrate into the inner bark, killing vital cells.

Management

Adequate control measures are not known for chestnut blight. Large sums of money have been spent on research to find ways to combat this disease, and extensive attempts at eradication have been tried with little success. Sanitation measures involving the removal and destruction of infected trees are expensive and unsatisfactory (Fig. 14.2). Managing the bark beetles by spraying with insecticides delays spread of the fungus, but spraying must be continued year after year. Resistant cultivars have not been developed. Chemicals injected into the sap stream give some control, but it is too early to determine their long-term effectiveness.

One hope in the battle against chestnut blight is the use of isolates of the causal fungus, *Endothia parasitica,* that have a low degree of virulence, i.e., they cause very little disease on a very susceptible host even under the most suitable environmental conditions. These *hypovirulent* (hypo = low) isolates can be used to infect existing chestnut sprouts and will compete with the highly virulent isolates that have been responsible for destroying trees. Not only would there be a competition for infection sites, but if the hypovirulent

Fig. 14.2. No really effective methods are available for managing chestnut blight, Dutch elm disease, or oak wilt. Dead or dying trees growing near houses or powerlines or along streets (such as these diseased elms) must be removed before they blow over and hurt someone or cause property damage. Oftentimes the trees must be cut down limb by limb. Cities and towns have spent millions of dollars to get rid of diseased trees.
Courtesy L. R. Schreiber

isolates fuse with the highly virulent isolates, then genetic material is transferred among the isolates and a highly virulent isolate may be transformed into a hypovirulent isolate. It is hoped that introduction of hypovirulent fungal isolates into forests will permit the mighty chestnut tree to thrive once again.

Dutch Elm Disease

Dutch elm disease, first reported in the United States in 1930, owes its name to the fact that it was first described in Holland. (There is not a tree known as Dutch elm.) The disease is caused by the fungus *Ceratocystis ulmi* (Buism.) C. Mor., which attacks and kills the American elm. This stately tree, which once lined the residental streets of most towns in the eastern and midwestern United States and inhabited the meadows and fence rows, has now largely disappeared. Aside from being deprived of the natural beauty of this shade tree, the cost of cutting down dead elm trees amounts to millions of dollars per year.

Symptoms

Trees suffering from Dutch elm disease show a variety of symptoms. Leaves become yellow, wilt, and turn brown. The leaves fall prematurely, and the branches die, causing the crowns to appear thin and sparse. Internally the outer sapwood turns brown. Many dead branches may appear on a tree. The trees may take several years to die, or they may die the same season they are infected.

Causal Agent and Disease Cycle

Although *Ceratocystis ulmi,* the causal agent of Dutch elm disease, produces three kinds of spores, the ones most commonly seen are produced on short black stalks, called *coremia* (Fig. 14.3), which are scarcely visible to the unaided eye. Each is about 1 mm high with an enlarged head bearing great numbers of egg-shaped, colorless spores (3 μm × 2 μm) embedded in a clear sticky fluid. The coremia grow in cracks, underneath loosened bark, and in the tunnels made by bark beetles (Fig. 14.4). If a beetle touches a coremium the spores stick to the insect. Contaminated beetles emerge from the feeding tunnels, fly to healthy trees, feed on small tender twigs, and unknowingly deposit the spores in the feeding wounds in the sapwood. Here the spores germinate and start new infections.

Usually there are two broods of bark beetles each year. The beetles reproduce in great numbers. As many as 25,000 beetles have emerged from 0.09 m² (1 ft²) of bark. The overwintering brood emerges in May or June and is chiefly responsible for spreading the fungus. The second brood emerges in August. After feeding, these beetles may fly more than 3 km to find a breeding place on dead or dying elms.

Once the fungal mycelium is growing inside the water-conducting vessels, it produces small, oval, colorless conidia; these are carried in the sap stream to other parts of the tree where they germinate. As a result of these multiple infections the tree usually dies.

Birds also carry the fungus long distances. To a limited extent water and wind also spread the spores, but bark beetles are the principal vectors. The fungus also spreads from diseased elms to nearby healthy elms by root grafts.

Management

As with chestnut blight, there are not satisfactory methods for managing Dutch elm disease. Refer to the discussion in Chestnut Blight section.

Oak Wilt

A third fungus that is quietly devastating our forests is *Ceratocystis fagacearum* (Bretz) Hunt, a close relative of the fungus that causes Dutch elm disease. At least 50 species and cultivars of oaks are susceptible. Oak wilt has caused the most damage around the Great Lakes, but it now occurs widely in the eastern United States and continues to spread.

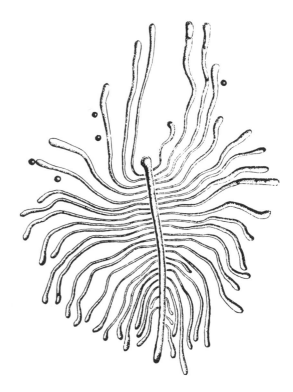

Fig. 14.3. **Spore-bearing stalks (coremia) of *C. ulmi*, the causal agent of Dutch elm disease. Magnified ×25.**
Courtesy R. K. Jones

Fig. 14.4. **Brood gallery, about 5 cm long, made by a female bark beetle between the bark and wood of an elm tree. The eggs are laid in the central tunnel. The eggs hatch and the feeding larvae open tunnels at right angles to the central gallery. The Dutch-elm-disease fungus grows in the tunnels and produces spores which stick to the adult beetles wherever they go, thus spreading the fungus.**
Courtesy USDA

Symptoms

Oak-wilt symptoms are most noticeable during the late spring or early summer. Red oaks may be killed in 3 weeks. In white oaks, symptoms are usually confined to a few branches, and infected trees may live several years. Leaves turn yellow or brown, dry up, and fall prematurely. A definite characteristic of oak wilt is the raising and cracking of the bark due to the pressure exerted by fungal mats growing between the bark and wood. Each mat consists of a central, black, furrowed, gelatinous sclerotium-like pad or cushion surrounded by mycelial growth. As the pad develops, it puts pressure on the bark, causing a bulge and forming a moist chamber, which is excellent for spore production. The continued growth of the pad eventually ruptures the bark. Since the mat produces an odor similar to fermenting apples, insects are attracted to the mats on which they feed. The spores stick to the insect's body and may be carried to other trees to start new infections.

Causal Agent and Disease Cycle

The oak-wilt fungus, *Ceratocystis fagacearum* overwinters as mycelium in infected trees and as fungal pads on dead trees. In the spring the pads produce great numbers of one-celled, hyaline, cylindrical conidia (ave. 13 μm $\times$ 3 μm) in the bark-beetle galleries (Fig. 14.5). The beetles feed on the pads, become contaminated, and may carry the spores long distances introducing the fungus into healthy trees through feeding wounds. The fungus spreads rapidly within the water-conducting tissues of the host causing the leaves to wilt. After the tree is killed, the fungus grows throughout the outer wood and into the roots. Adjoining oaks can become infected through root grafts. Pressure from mycelial pads formed under the bark of dead trees force it to crack. Insects enter the cracks and feed on the pads.

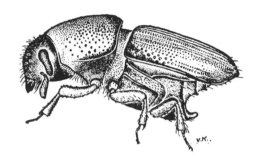

Fig. 14.5. Mature bark beetle, *Scolytus* sp., about 3 mm long. Bark beetles spread Chestnut blight, Dutch elm disease and oak wilt fungi.
Courtesy N.C. Agric. Extension Service

Management

See discussion on management in Chestnut Blight section.

Annosus Root and Butt Rot

Annosus root and butt rot is common throughout coniferous stands of the North Temperate Zone. Hardwoods may be attacked, but damage is usually of little importance. In the southern United States, the disease is most serious in pine plantations on sandy soils low in organic matter. All species of southern pine are susceptible. Slash and loblolly pine plantations are often severely affected.

Symptoms

The disease gains entry into plantations by spore infection of freshly cut stumps exposed during thinning operations. The fungus then spreads from infected stumps to nearby trees by growth along the roots to points of root contact. Residual trees usually begin to die within a few years after thinning.

The *sporophores* or conks (fruiting bodies) are generally found at ground line or in the root crotch, partially hidden by ground litter (Fig. 14.6). Pines in initial stages of the disease usually have fewer leaves than normal; however, white pines with full crowns may have extensive butt and root decay. Occasionally trees may die rapidly with a sudden red discoloration of a nearly full crown. Diseased trees are often found in groups or circular areas in the forest. Symptoms of the disease may include the pink to violet stain of new decay, narrow white pockets and scattered black flecks of decay in the wood, and the yellow stringy rot of the late stages.

Causal Agent and Disease Development

Annosus root rot is caused by *Heterobasidion annosum* (Fr.) Bref. The fungus forms thin mycelial mats on infected trees between the bark and wood and small dirty-white outgrowths of mycelium on the undersides of the roots. The grey-brown perennial conks or sporophores are flat to bracket-like and measure 1–15 cm × 2–25 cm × 0.5–7 cm. The colorless microscopic basidiospores (3.5–5 μm × 3.4 μm) look like tiny teardrops. They are produced in great numbers in the conks and are blown by the wind to infect recently cut stumps.

Decay progresses from the infected stump into the roots. Roots of adjacent trees are infected by contact with decayed roots and through root grafts. After stump invasion, it usually takes from 4 to 10 years for symptoms to appear in nearby trees. Although most infection occurs through wounds, the fungus can penetrate nonwounded bark. The fungus survives in a stump for many years but does not exist long in the soil. Once established in a stand, the fungus remains indefinitely.

Fig. 14.6. Conks or sporophores (about one-half natural size) growing at base of pine tree infected with annosus root and butt rot.
Courtesy USDA

Annosus Root and Butt Rot   **219**

Management

To minimize the chances of infection by the causal fungus, growers should avoid planting pine trees on soils low in organic matter and, if possible, eliminate thinning on disease-free sites. Immediately after trees are cut, technical grade granular borax should be applied to stump surfaces with a large "salt shaker." This fungicide serves as a chemical barrier to any fungal spores that happen to fall on the stump. Borax is effective, safe, easy to apply, inexpensive, and not harmful to the environment.

In infected stands, tree stumps can be sprayed immediately after cutting with a spore suspension of a competing fungus, *Peniophora gigantea* (Fr.) Mass., which prevents infection. This use of a competing, nonpathogenic fungus to prevent infection by a pathogenic one is another example of biological control.

Fusiform Rust

Fusiform rust is one of the most destructive diseases of forest trees in the United States. The disease occurs from Maryland to Florida and westward to Texas and Arkansas. Serious economic losses occur in nurseries, seed orchards, young plantations, and natural stands. During the past 50 years the disease has increased in intensity and frequency; in 1978, stumpage losses were estimated at $28 million annually. Loblolly and slash pines are very susceptible; longleaf pine is moderately resistant, and shortleaf pine is highly resistant. Damage to oaks, the alternate host for the fungus, is negligible.

Symptoms

The most characteristic symptom of fusiform rust is a tapered, spindle- or fusiform-shaped gall that forms on branches and stems of young pines. These galls often encircle the stem or branch causing it to die. In early spring these swellings appear yellow to orange as the fungus produces many spores. The fungus stimulates excessive branches in young trees. They may develop so many branches that they resemble brooms. As host tissue is killed, older stem cankers—sometimes reaching 30 cm in length—become flat, sunken, and soaked with pitch. Mortality is heaviest on trees less than 10 years old, but the galls and resulting cankers deform older trees, reducing growth and weakening the stems so that they are easily broken during wind or ice storms.

Causal Agent and Disease Cycle

Fusiform rust is caused by *Cronartium fusiforme* Hedge & Hunt ex Cumm. It has two hosts—oak and pine—and five kinds of spores. Although the fungus takes 2 years or more to complete its life cycle, most of the active growth occurs in the early spring. From late February to early April, the fusiform galls produce enormous numbers of orange aeciospores. These are blown to the young oak leaves that are emerging about this time where they germinate and penetrate through the stomates. Only very young oak leaves, less than 2 weeks old, can be infected.

Within 7–10 days after infection, the fungus produces small pustules containing red uredospores on the lower surface of oak leaves. The uredospores can also infect oak leaves and are called repeating spores. After about a week, brown, hairlike structures (telia) appear on the underside of infected oak leaves. The telia contain the teliospores. During favorable conditions (15°–27°C and 97–100% RH), the teliospores germinate, each producing four delicate, short-lived, basidiospores, which are blown to susceptible pine hosts where infection occurs on cotyledons, needles, or succulent tissue.

Most infection occurs at night and in the early morning following a day or evening of rain. Once established in host tissue, the fungus stimulates excessive growth in infected tissues resulting in gall formation. If the disease does not kill the pine host during the first few months after infection, the newly formed galls exude yellowish drops of fluid full of pycniospores. Aeciospores are produced on the galls the next spring, completing the life cycle. The fusiform galls produce aeciospores successively for many years. Old galls or cankers are often invaded by insects and wood-rotting fungi that reduce a tree's value. Such weakened trees break easily or snap off during windstorms, often falling or lodging against other trees. This makes cutting operations more difficult and expensive.

In oaks, only the leaves are infected by the fusiform-rust fungus, and the fungus is killed when the temperature remains over 30°C for several days. The oaks are free of disease in winter when they cast their leaves. In pines the fungal mycelium lives in infected tissue throughout the year.

Environmental Effects

Favorable conditions for *Cronartium fusiforme* infection on pine occur when the RH remains at or near 100% and temperature ranges between 15° and 27°C for as few as 12 hr; such conditions are common on many spring nights in the southern United States. When rain, fog, or overcast skies extend the time these conditions occur, during a period that coincides with flushes of new growth on pine, infection rates can reach epidemic proportions.

Management

Fusiform rust can be managed in nurseries by spraying with recommended fungicides. If adequate weather forecasts are available, the nursery can be sprayed prior to periods favorable for rust infection. Usually three to four applications are necessary.

Infected seedlings should not be used. Some diseased seedlings may not have visible galls. If a given bed contains more than 1% of visibly galled seedlings, it may be desirable to discard the entire bed.

Neither slash nor loblolly pines should be planted on sites with histories of severe rust infection or where there is an abundance of oaks nearby. Plant longleaf or shortleaf pine instead.

The most promising method for managing fusiform rust in future years is the use of seed from resistant trees or hybrid seed produced by crossing shortleaf pine with either loblolly or slash pine. Seed orchards of rust-resistant clones of loblolly and slash pines are now established. As seed from these orchards become available, they should be used in areas of high rust hazard.

Little Leaf of Pine

Symptoms

Little-leaf disease is most evident in older pine trees, rarely attacking stands less than 20 years old. The disease is caused by a malfunctioning of the root system resulting from a combination of biological and physical factors. The fungus, *Phytophthora cinnamomi* Rands, attacks and kills the root tips. When conditions of moisture, fertility, and drainage are not good, such trees are not able to rapidly replace the destroyed root tips. Trees on good sites with adequate moisture, fertility, and drainage are reportedly also attacked by the fungus, but their vigor is such that they quickly overcome the disease by producing new root tips.

The disease usually progresses rather slowly. Some trees may live 15 or more years after the appearance of initial symptoms. In general, trees live only 5–6 years after attack, but they may die in as little as 1 year. Symptoms are those typical of trees stressed by a malfunctioning root system. In the early stage of the disease, the foliage may turn yellow-green and the current year's needles are shorter than normal. In later stages, diseased trees have sparsely foliated crowns with shortened needles and dead branches. Abundant foliage sprouting on the boles of infected shortleaf trees is common.

Causal Agent and Disease Cycle

Phytophthora cinnamoni, which causes little leaf of pine, is similar to other species of this fungus that cause various root and stem rots. For a discussion of these *Phytophthora* species, see Chapter 12, Black Shank section.

Management

At present, effective management of little-leaf disease is not available. Losses can be minimized by salvage cuttings and by using loblolly pine and hardwoods in reforestation plantings.

However, research since 1970 on some fungi that grow in association with the roots of many trees offers hope for biological management of little leaf of pine in the future. Some of these fungi—called *mycorrhizae*—grow inside roots; these are called *endomycorrhizae*. Others are attached to the outside of roots and form a cover or mantle around the roots; these are called *ectomycorrhizae*. Oddly enough, pine seedlings supporting ectomycorrhizae on their roots are resistant to infection by the fungus that causes little leaf of pine, whereas seedlings with nonmycorrhizal roots are heavily infected by

Phytophthora cinnamomi under similar conditions. Perhaps someday soon roots of pine seedlings can be inoculated with harmless mycorrhizal fungi and thus be protected against little leaf. If this practice proves feasible and effective, it would be an excellent example of biological management of a forest tree disease and might be applicable to the management of root diseases in other crops.

Wood Decay in Buildings

Wood is one of man's most versatile and durable building materials. When properly used, it will give long and satisfactory service. If you ignore a few simple rules, however, wood-decay fungi can attack wood. Management then can be very costly. Termites and some other insects also cause damage to buildings, but management and prevention of insect attack will not be discussed here.

Causal Agents and Symptoms

Two groups of fungi cause wood decay. Sap-stain and mold fungi discolor wood but do not seriously weaken it. Sap-stain, surface molds, and mildews can be black, blue, grey, green, orange or pink. They all grow under the same conditions and are sometimes found together, but not always. The second group, the decay fungi, attack wood that remains moist because of rain or groundwater seepage, leaky plumbing, or condensation. Such decay, commonly known as wet-rot, dry-rot, or brown cubical rot, does not extend beyond the moist areas.

Occasionally, wood is attacked by a fungus that first decays moist wood and then conducts water to normally dry wood. Consequently, the fungus advances more rapidly than ordinary decay fungi. This fungus is *Poria incrassata* (B & C) Burt., the water-conducting or dry-rot fungus. It can destroy the wood in a building within 2 years. Although this does not occur very often, it is possible for this fungus to grow where ordinary decay fungi will not grow.

Below 5°C decay of wood is slow. When its moisture content is below 20%, wood will not decay. Wood that is either properly treated with a good preservative or comes from the heartwood of durable tree species will not decay even when its moisture content is above 20%.

Poria species can conduct or pump water several meters from soil or other sources, sometimes through brick or concrete, by means of thick rootlike strands (rhizomorphs) up to 1 cm or more wide. These strands, which are often hidden behind wooden structures, distinguish the water-conducting fungus from ordinary wood-decaying fungi. If wood decay is caused by a *Poria* species, greater care is required to control attack than for ordinary decay fungi. If the cause of decay is not clear, seek advice from the local county extension office, pest control company, or building inspector.

Darkening of wood, along with cracking and splitting across the grain, are the most common symptoms of more advanced decay in structural timbers. Finally, the wood will appear quite dry and crumbly.

Management

Arresting Wood Decay

The first step in halting the spread of wood decay in buildings is to find and correct all sources of excessive moisture. Some common sources included the following:

- Poor drainage or poor foundation waterproofing around the building so that water runs into or through the foundation
- Inadequate ventilation in crawl spaces or other enclosed areas of construction
- Leaky plumbing
- Untreated wood in contact with soil, particularly in earth-filled porches

If possible, eliminate excessive moisture, remove all obviously decayed wood, and replace it with dry, sound lumber. It is not necessary to remove sound wood around the decayed area. If excessive moisture cannot be fully eliminated, pressure-treated wood should be used to replace the decayed wood and surrounding wood for a distance of at least 0.7 m beyond any evidence of decay.

Localized decay in wood structures such as joists, porch floors, or bases of columns can be arrested by flooding the wood surface with a water-repellent preservative, providing the wood can be dried out before treatment.

Helpful Hints to Prevent Wood Decay

To prevent wood decay, it is necessary to keep untreated wood dry. The following practices are recommended (Fig. 14.7):

- Provide adequate drainage and foundation waterproofing to prevent water accumulations under or near the building.
- Cover at least 70% of the crawl space floor in basementless houses with polyethylene (4 or 6 mil), or place polyethylene directly over or preferably directly under concrete slabs.
- Provide adequate ventilation in crawl spaces and attics.
- Install a moisture barrier between sills and foundations.
- Avoid all forms of construction that will trap moisture.
- Insulate any pipes likely to cause condensation of water onto wood.
- Keep gutters clear of leaves and other plant debris.
- Avoid discharge of water or water vapor (e.g., from dryers or air conditioners) into crawl spaces.
- Keep untreated wood well above the soil.
- Remove all forms used in pouring concrete and wood debris from under a building.
- Keep caulking and paint around joints well maintained.

Crawl spaces should be cross-ventilated and contain at least 0.1 m^2 of clear vent space per 15 m^2 of crawl space. Vents should be kept open and should not be blocked by leaves, shrubs, wood piles, or trash. Attics should contain 0.1 m^2 of vent space per 15 m^2 or the ceiling below; however, when more than 50% of the vent space is near the top of the gable, 0.09 m^2 space per 27 m^2 of the ceiling below is sufficient.

YES

Roof ridge straight

At least 18-Inch roof overhang at eaves and gables

Gutters tight and free of debris

Shingles cover all roof decking

Attic vents

Porch sloped away from house

Vents for crawl space

Ground sloped away from house

Porch below wood members

More than 18-inch crawl space

Planter separated from foundation

Porch separated from foundation

No peeling of paint.

Downspouts intact

Caulking maintained

NO

Gutters leaking

Roof overhang less than 18 inches

Roof ridge sagging

Roof decking exposed

Unvented attic

No slope on porch

Paint peeling.

Planter against foundation

Porch above wood members

Caulking gone

Downspouts broken

Porch against house

Unvented crawl space

Water puddles next to house

Less than 18-inch crawl space

Fig. 14.7. **Proper and improper house construction. The bottom sketch shows features that invite decay.**
Courtesy USDA

Foundation walls supporting wood frame construction should be at least 20 cm above outside finish grade. In buildings without a basement, a crawl space should have at least 45-cm clearance under wood sills and joists. Wood siding should be at least 15 cm above the outside finished level of soil. Earth-filled porches should be avoided unless the earth is at least 20 cm away from wood construction.

Buildings should be inspected regularly for the warning signs of decay and for the conditions that may lead to decay. Ideally, inspections should be made twice yearly, in mid-summer and mid-winter.

Wood Preservation

The most efficient way to prevent or control decay when wood is used where it will get wet is to treat it with a chemical. Fence posts, pilings, railroad ties, and utility poles are usually pressure-treated with *creosote*. In the southeastern United States, untreated fence posts will often rot in 3–4 years, whereas properly treated posts will last 20 years. Creosote-treated posts are usually available at most building material companies or farm supply outlets.

The odor and appearance of wood treated with creosote makes it undesirable for some uses; therefore, other oil-soluble preservatives are used for many products. *Pentachlorophenol* and *copper naphthenate* are commonly used to treat wood used for benches, doors, interior framing, fence posts, floor joists, sundecks, windows, and other wood products exposed to the weather. Some uses of creosote and pentachlorophenol may be eliminated, but other chemicals are available or are being developed.

For wood used above ground and where the decay hazard is relatively small, dip and brush-on treatments provide adequate protection. These dip or brush-on treatments should be used on wood with a moisture content of less than 12%. Wood that has been treated with certain chemical preservatives is difficult to paint properly.

15

Bacteria

Bacteria—our most useful servants and most destructive masters—are essential to human existence. Bacteria produce useful food and chemicals; increase the fertility of soil, fix nitrogen from the air, aid in digestion, decompose the complex bodies of dead plants and animals, and salvage the parts so that they may be used again. But, as causes of disease, bacteria have caused more deaths than all wars. Bacteria have defeated armies and destroyed nations. Yet most bacteria are beneficial, not harmful.

Discoveries about the nature and activities of bacteria are among the most important scientific accomplishments of modern times. Anton von Leeuwenhoek, a Dutch lens grinder, first saw bacteria about 1683 in a crude microscope he built himself. In 1876 Robert Koch, a German country doctor, showed that a specific bacterium was the cause of a specific disease—anthrax—in animals. We now know that bacteria cause numerous diseases in human beings including typhoid fever, tuberculosis, diphtheria, lockjaw, and syphilis. A few years later, in 1882, Louis Pasteur identified bacteria as the cause of fermentation. Many plant parts, when deprived of free oxygen still continue to give off carbon dioxide and often produce alcohol and other compounds. This is called *anaerobic respiration* or *fermentation.* The theory that living agents cause fermentation became the basis of modern bacteriology and laid the foundation for aseptic surgery, the control of epidemics, preservation of foods, purification of water and milk (heating milk to destroy bacteria is called *pasteurization*), and sewage disposal.

As a result of the work by Koch, Pasteur, and others, the *germ theory of disease* became widely accepted. This theory states that diseases are caused by living agents (*germs*). Thomas Burrill, a professor of botany at the University of Illinois, first discovered in 1881 that bacteria caused plant diseases when he showed that these microscopic germs caused fire blight of pear and apple.

Size and Structure

Bacteria are microscopic, unicellular organisms, often known as germs or microbes. They have no roots, stems, or leaves and contain no chlorophyll. Some are so small that 25,000 laid end for end would be only 2.5 cm (1 in.) long. A bacterium consists of only one cell that varies in size from 0.3 to 6 μm in length (see Fig. 2.1). Thousands of bacteria might be contained in a single drop of water and not be crowded at all for space. A milliliter (20–30 drops) of sour milk contains millions of bacterial cells.

Bacteria are classified into three general types, depending on their shape (Fig. 15.1): (1) spherical or globular (*coccus*), (2) rod or cylindrical (*bacillus*), and (3) spiral (*spirillum*). Only non-spore-forming bacilli (about 1 μm in length) are plant pathogens. The cells of the different types may be held together as pairs, long or short chains, long spirals, or clusters.

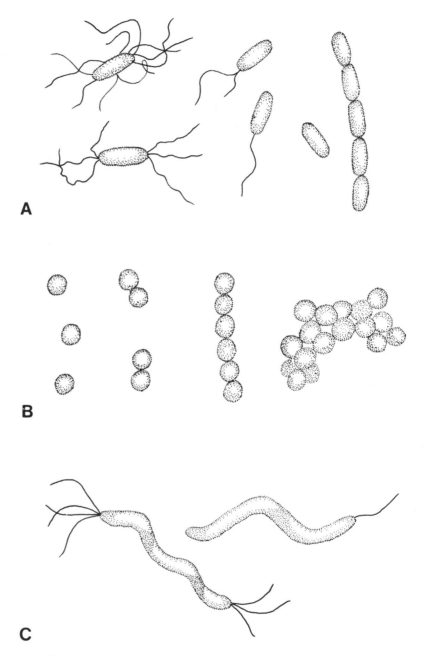

A

B

C

Fig. 15.1. **Types of bacteria, based on their shape, magnified about 900 times. A—Rod. B—Spherical. C—Spiral.**
Drawings by Pi-Yu Huang

Each bacterium consists of a cell wall, cytoplasm, and nuclear material (Fig. 15.2). Many bacteria secrete a sticky or gummy material called *slime,* which accumulates around the cells and helps hold them together.

Some bacteria have long slender, hairlike, coiled appendages called *flagella.* By their lashing movement, the flagella propel a bacterial cell through water or nutrient solutions. Bacterial movement may be straight, wiggling, or tumbling. Spiral forms may whirl like a ship's propeller. Some bacteria have only one flagellum attached to one end of the cell; others have many flagella distributed over the cell surface.

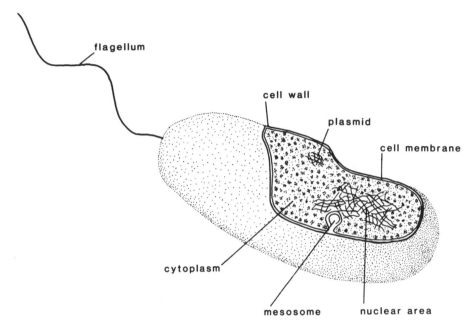

Fig. 15.2. **Diagram of a bacterial cell with wall cut away to show anatomy.**
Drawing by Pi-Yu Huang

Identifying Bacteria

Many plant-pathogenic nematodes and fungi are big enough and different enough in size, shape, and/or color that they often can be identified by just looking at them. But not so with bacteria. Most plant-pathogenic bacteria are short rods which you cannot tell apart even with the most efficient light microscope. However, it is essential to identify different bacterial species to study them and take proper measures to manage them. Since bacteria are so similar and cannot be identified by shape, size, and/or color only, scientists use other characteristics to differentiate bacteria; these include their chemical composition, enzymatic action, serological specificity, pathogenicity to plants, and susceptibility to certain bacterial viruses called *bacteriophages.*

Chemical composition can be detected with specific staining techniques. Gram's method, developed in 1884 by Christian Gram, involves the use of crystal violet and iodine to determine if bacteria will absorb these chemicals and turn dark (gram positive). Most plant-pathogenic bacteria do not retain the crystal-violet stain and are called gram negative.

Enzymatic action is determined by recording the substances that bacteria can use for food and also the substances produced when bacteria grow on certain substrates. For example, some bacteria may change nitrates to nitrites; others produce hydrogen sulfide, ammonia, or many other substances. Determinations of the minimum, optimum, and maximum temperature and pH for growth, and the temperature at which bacteria are killed are also used to identify bacterial species.

When an unknown bacterium has been put through all these tests (which may total 50 or more), it can be identified by comparing its reactions to those listed in tables available for the known species of bacteria. The unknown is considered to be the same as the listed species that gives the same reactions, especially when it can be shown to attack various species and cultivars of host plants.

Growth and Reproduction

Bacteria multiply by dividing! Each bacterium grows until it reaches a certain size, then the cell becomes pinched in the middle and separates into two equal parts or daughter cells. This process is called *binary fission*. Each daughter cell quickly enlarges, then divides again. Some bacteria divide every 20 to 30 min. Such a rate of division can be maintained for only as long as sufficient nutrients are present. Crowding, depletion of food and water, accumulation of poisonous waste products, and changes in the pH of the environment can limit bacterial populations.

The ability of bacteria to produce tremendous numbers of individual cells in a short time and their production of toxins and enzymes that harm the host explain why bacteria are so destructive to animals and plants. A single bacterial cell is harmless, but severe damage occurs when a great many develop in wounds, or other natural openings.

Bacteria do not contain chlorophyll and cannot manufacture their own food.

Many bacteria grow on the dead remains or the products of plant and animal life and do not cause disease; they are *saprophytes*. Some species are parasitic; they attack the living cells of plants or animals and secure their food from them. Some parasitic bacteria such as the root nodule bacteria are useful and some cause diseases. Many of the saprophytic bacteria are beneficial. Bacteria that can grow on a living or nonliving host are called *facultative parasites*. Many of the bacteria that cause plant diseases are in this group.

Bacteria produce many enzymes that digest or change complex food materials into simpler compounds and also help build simple compounds into larger compounds. Certain bacteria are able to synthesize all compounds necessary for growth from sugars and essential elements. Others must obtain certain growth factors, vitamins, and/or amino acids from the environment.

Some bacteria require free oxygen for growth; others can grow in the absence of oxygen. Some can do both. In addition, bacteria require water and suitable temperature, osmotic concentration, and pH for growth.

Many saprophytic bacteria are capable of producing profound chemical changes in the substance on which they grow. Decay of plant and animal materials and processes known as fermentation are changes of this type. Decay includes the decomposition of organic material into its constitutents or to such simple compounds as water, carbon dioxide, ammonia, and hydrogen sulfide. Many bacterial species may be involved in the decay of a dead plant or animal, each breaking down some special compounds into simpler substances. Other species carry the destruction a step further, until it is completed. One kind of bacteria could not completely decay the carcass of an animal or the body of a plant.

Decay processes are important to man from at least two standpoints. They prevent the accumulation of organic matter—both plant and animal—on the earth, and they result in the formation of simple compounds or free elements that are returned to the soil, water, and air to be used again by plants.

It must be remembered that the cell walls and membranes of bacteria are not hard and stiff like steel, but are soft and pliable. Water itself and some compounds dissolved in water easily pass through the cell walls into the cytoplasm where they are transformed by the bacterial enzymes to supply food and energy. Materials also can move outward from the bacterial cell. For example, toxins formed inside bacteria may leak through the bacterial walls and react with plant juices to kill the tissues.

Distribution and Dispersal

Bacteria occur everywhere—in soil, water, air, food, dust, the oceans. They are found on and in plants and animals. Bacteria enter plants through natural openings (stomates, lenticels, hydathodes) and wounds. Once inside, the bacteria are carried in the sap stream or swim in the water between the cells.

Bacteria cannot spread long distances on their own but are easily spread in many ways. Soil, refuse from a previous crop, compost, diseased transplants, contaminated seed, splashing rain, insects, nematodes, birds, man, water, and wind—all may carry bacteria to healthy plants, from field to field, or over long distances from country to country.

Managing Plant-Pathogenic Bacteria

Bacterial diseases of plants are managed by use of resistant cultivars, crop rotation, sanitation, disease-free seed and transplants, and chemical sprays or antibiotics. The most effective antibiotics against bacteria are obtained from bacteria and fungi. Bacterial diseases can also be managed by reducing wounding or damage to plant parts.

Bacteria can cause rot or decay when excessive moisture is present. In addition, many bacteria require warm temperatures to grow and multiply.

Therefore, bacterial rots of harvested grains (corn, oats, rice, wheat) and other crops are controlled by first drying them and then storing under very dry conditions. Harvested vegetables and fruits are stored at temperatures below 10°C.

Bacteria may also be attacked by viruses called bacteriophages. These bacteriophages attack the bacterial cell and eventually kill it. In the future bacteriophages may become a useful tool in the management of bacterial diseases.

16

Diseases Caused by Bacteria and Mycoplasma

When bacteria fill the water-conducting cells of a plant and kill them by means of toxins, the plant suffers a *wilt* or *blight* disease. If the bacteria kill leaf cells, they may cause a *leaf spot.* When they infect fruits, tubers, or roots, they cause *soft rots,* which result in a slimy, foul-smelling mass. Other bacteria stimulate tissue growth to form *galls* or tumors that resemble cancerous tissues of animals. In addition to killing plant tissues, bacteria can also stunt the growth of plants, change their color from green to brown or black, or cause distortion of leaves, stems, flowers, or fruits.

Several typical bacterial diseases of plants are discussed in this chapter. These are summarized in Table 16.1.

Mycoplasmas, organisms smaller than most bacteria, cause a group of diseases distinguished by stunting, chlorosis, yellowing of leaves, changing of flowers to leaflike structures, production of numerous leaves, or some combination of these symptoms. A discussion of mycoplasma diseases appears in this chapter.

Table 16.1. Several Important Diseases Caused by Bacteria

Disease	Pathogen	Hosts
Wilt or Granville wilt	*Pseudomonas solanacearum*	Tobacco, tomato, potato, eggplant, and others
Fire blight	*Erwinia amylovora*	Apple, pear
Soft rot, Barn rot, Storage rot	*Erwinia carotovora* subsp. *carotovora*	Potato, cabbage, lettuce, carrot, tobacco
Wilt	*Erwinia tracheiphila*	Cucumber
Canker	*Corynebacterium michiganense* pv. *michiganense*	Tomato
Spot or Canker	*Xanthomonas campestris* pv. *pruni*	Peach
Crown gall	*Agrobacterium tumefaciens*	Fruit trees, woody ornamentals

Bacterial Wilt

The wilt disease caused by the soilborne bacterium *Pseudomonas solanacearum* Smith, also called Granville wilt, is one of the most important and widespread bacterial plant diseases. During the past 50 years the total losses from this disease on all crops have been worth hundreds of millions of dollars. Bacterial wilt has been reported from most countries in the temperate and semitropical zones of the world. It is an annual threat on sandy soils to such crops as banana, eggplant, peanut, pepper, potato, tobacco, and tomato. Certain ornamental plants—including dahlia, marigold, nasturtium—and many weeds are hosts to this wilt bacterium.

Symptoms

Typical external symptoms of bacterial wilt include stunting, yellowing, and wilting of the aboveground parts (Fig. 16.1). Leaf *epinasty* and adventitious roots may appear on certain host plants. The vascular tissue invaded by *P. solanacearum* usually becomes dark brown. Many roots of infected plants decay, becoming dark brown to black in color.

If the stem of a diseased plant is cut in cross section, an exudate of bacterial slime in the form of tiny droplets may become evident on the cut surface. The presence of the wilt pathogen also can be demonstrated by floating a razor-thin slice of a longitudinal section containing vascular tissue from a diseased stem in a container of clear water. Within a few minutes, fine milky-white strands composed of masses of bacteria will stream from the margin of the tissue.

Fig. 16.1. Left: Healthy tomato plant. *Right:* Stunted, wilted plant with yellow leaves suffering from bacterial wilt.
Courtesy S. F. Jenkins

Causal Agent

The wilt bacterium, *Pseudomonas solanacearum* Smith, is a gram-negative, non-spore-forming rod, about 0.5 mm × 1.5 mm, with a single polar flagellum. The bacterium is aerobic, reduces nitrate, and produces ammonia. Colonies on solid media are small, irregularly round, white in reflected light, and tan in transmitted light. In certain simple synthetic media, the bacterium can utilize inorganic sources of nitrogen and most of the common carbon sources. The organism is sensitive to desiccation. On plant-extract media and on potato plugs a characteristic brown pigment is formed. Rapid growth *in vitro* is favored by relatively high temperatures (30°–35°C) and a neutral to slightly acid medium (pH 6.7). Unless special procedures are followed, such as storage under sterile mineral oil or sterile distilled water, pathogenicity and viability may be lost rapidly in culture.

Strains of *P. solanacearum* differing in cultural and biochemical characteristics and in levels of virulence have been reported. Bacteriophages (viruses) that attack and kill the bacterium have been found.

Disease Cycle

Bacterial-wilt infection usually occurs below ground through wounds in the root system. The root wounds made by the invading larvae of the root-knot nematode provide excellent places for the wilt bacterium to enter the roots. Apparently, root-knot-infected tissue acts as an excellent food source for the bacterium. The organism moves upward in the vessels from secondary roots into larger roots and the stem. Bacterial cavities may be formed in local areas in the phloem, cortex, or pith. When the number of vessels blocked by masses of bacterial cells, and probably also by gums and slimes (Fig. 16.2) of bacterial and plant origin, become great enough, the infected plant suffers irreversible wilting followed by death. Toxic materials such as growth inhibitors and hormones probably play an important role in development of wilt symptoms and death of host tissues.

There are many known cases in which bacterial wilt has caused heavy losses of tobacco and bananas in newly cleared land. Apparently the bacterium is a normal inhabitant of many soils. Although the bacterium survives longer in some soils than others, for all practical purposes a field will remain contaminated indefinitely once it has become infested with the wilt organism. Most of the bacteria are found in the plow layer (top 30 cm of soil).

The wilt pathogen can be spread in seedlings used as transplants, in potato seed tubers, in banana suckers, by various cultural practices involving spread of soil or plant debris, and by movement of soil by surface water. Dispersal by insects is of minor significance, and spread in true seed has not been conclusively demonstrated. In general, the rate of dispersal in a given area is relatively slow.

Disease development is favored by relatively high temperatures and high soil moisture levels. As a result, seasonal fluctuations in disease severity may be very great. In greenhouse experiments an interrelationship between disease severity, day length, temperature and host nutrition has been demonstrated.

Fig. 16.2. The bacterial-wilt pathogen, *Pseudomonas solanacearum,* magnified 1800 times. A—Cells and slime of a highly pathogenic strain. B—Slime-free cells of a weakly pathogenic strain. C—Strands and bubbles of slime isolated from the filtrate of a highly pathogenic strain grown in culture. D—Cells and slime from the same strain from a diseased plant.
Courtesy A. Kelman

The bacterium occurs throughout the world in many different soil types ranging from sandy to heavy clay and over a wide range in pH levels. In a given field the disease is usually more severe in localized areas, often associated with inadequate drainage. Soils that dry out quickly are unfavorable for bacterial wilt, as are soils high in organic matter that permit growth of organisms antagonistic to the wilt bacterium.

Management

Successful management of bacterial wilt requires an integrated management program using every sensible measure possible. Root-knot nematode management, discussed in Chapter 9, is necessary. Only healthy transplants and wilt-resistant cultivars of tomato and tobacco, where available, should be used. Crop rotation is recommended. Seedbed and greenhouse soils should be sterilized before susceptible crops are planted.

Fire Blight of Apple and Pear

When Koch, Pasteur, and others first proposed the germ theory of disease, they based their conclusions on observations of animal diseases. A few years later, in 1881, Thomas Burrill of the University of Illinois showed that bacteria caused fire blight of apple and pear, the first plant disease attributed to a bacterium. This discovery caused much argument, and it took 20 years or more to convince some scientists that bacteria caused plant diseases. Fire blight also was one of the earliest diseases demonstrated to have an insect vector.

Fire blight is still one of the most important diseases of apples and pears. In certain areas, including the coastal plains of the eastern United States, commercial production of pears, even if other diseases are absent, is impossible because of this disease. Many ornamental trees and shrubs, especially those in the rose family, are susceptible to fire blight. Their attractiveness is decreased by the ravages of this disease, and in addition they serve as an inoculum source for commercial orchards nearby.

Symptoms

Fire blight often appears suddenly in the spring, killing the blossoms, shoots, and fruit spurs on apple, crabapple, hawthorn, pear, and pyracantha. The flowers and leaves become watersoaked, then shrivel rapidly, and finally turn brown and then black as though scorched by fire. The symptoms spread rapidly to nearby twigs. Branches also become blighted and turn black; the dead, curled, black leaves clinging to blighted branches contrast starkly with the healthy green of nearby parts of the tree. Watersprouts ("suckers") usually wilt from the tip downward. The bark cracks, turns brown-black, hard and shrunken. Fruit set of blighted trees is reduced. Spurs, branches, and sometimes whole trees are killed in 1 year.

Under humid conditions, droplets of milky, sticky ooze appear on the surface of blighted twigs or branches. The ooze turns brown as it hardens and dries. Or the droplets may run together to form large drops, which drip onto and contaminate other parts of the tree.

Rot-producing fungi often invade blighted twigs and branches, which makes fruit-rot and leaf-spot management more difficult.

Causal Agent and Disease Cycle

Erwinia amylovora, (Burr.) Winslow *et al.*, the fire-blight pathogen, is rod shaped and about 0.6 μm × 1.5 μm. It is motile, possessing numerous flagella attached over its entire surface. It grows with or without free oxygen. Cultures often lose their virulence when maintained on artificial media.

The occurrence and severity of fire blight varies from season to season and orchard to orchard. The disease starts in the spring from bacteria that survive in holdover cankers. Rain spreads the bacteria to the blooms and increases the moisture level in the flowers, twigs, and leaves so that the bacteria can infect, often directly through the stomates, and grow in them. Strong winds during rains blow the bacteria into wounds and cracks on twigs or branches.

Temperatures between 20° and 28°C are most favorable for growth of fire-blight bacteria and disease development. Young, succulent tissues full of water are most susceptible. Aphids spread the bacteria, especially on watersprouts. Pollinating insects carry the bacteria from infected to healthy blooms, from tree to tree, and orchard to orchard, frequently during warm, sunny weather following rainy periods.

Management

A simple, practical, and fully effective management program for fire blight has not been developed. However, several practices can help check fire blight.

Holdover cankers and blighted shoots should be eliminated. Cut off branches with cankers during the winter. Another way to reduce the number of cankers on large branches and trunks of trees is to rub off all watersprouts during the spring and summer. Keep such shoots and spurs from developing on main branches. The source of blight for commercial orchards is often maintained in neglected pear and apple trees in yards, gardens, and home orchards. Such uncared-for trees in the near vicinity of commercial orchards should be removed.

Blossom blight infection can be greatly reduced by spraying with streptomycin during bloom, at rates to give 60–100 ppm streptomycin. Applications should be made at 5-day intervals through bloom. The 100-ppm rate is more effective than the 60-ppm rate. Apply the first spray in the late pink stage before many of the center blooms open. Leaves and blooms should be thoroughly sprayed, but overspraying increases leaf injury and should be avoided. Perfect control of blight should not be expected from bloom spraying because it is impossible to cover all the parts of the blooms and shoots that develop during rainy periods when blight infection takes place.

Postblossom spraying can help control twig and shoot blight. Streptomycin at 50 ppm is now registered in many areas for this use in postblossom sprays to be applied up to 50 days before harvest.

Trees subject to fire blight should not be overfertilized or stimulated in other ways to make extremely vigorous growth during and just after the bloom period. Blight is often less serious in sod orchards than in clean cultivated ones.

If noticed early in their development, blighted shoots may be broken off 25 cm below any sign of discoloration of the bark. Breaking, rather than cutting infected shoots results in less chance of spreading the bacteria. This procedure is practical only with young trees. Apple and pear cultivars that are extremely susceptible to fire blight should not be planted. All cultivars can be attacked by fire blight at times, but the more susceptible ones are Golden Delicious, Grimes, Jonathan, Yellow Transparent, Lodi, and Stayman. Some pear cultivars, such as Bartlett, are so susceptible to fire blight that they usually are killed if grown in the eastern United States. With other cultivars (Kieffer and Orient), the degree of injury from blight is usually less severe. Magness and Moonglow are two relatively resistant pear cultivars.

Soft Rots

Bacterial soft rot is one of the most destructive diseases of vegetables in storage and transit. It occurs most commonly on those vegetables that have fleshy storage tissues or fruits such as cabbage, carrot, celery, cucumber, eggplant, lettuce, onion, potato, radish, squash, sweet corn, sweet potato, and tomato. Hyacinth and iris bulbs are also affected. The soft-rot bacterium causes a disease on tobacco in the field called *hollow stalk* and a *barn rot* when the tobacco is being cured.

Soft rots occur all over the world and can cause serious losses during harvest, on the way to market, and especially in storage. Nearly all fresh vegetables may start to rot within a few hours in storage or during marketing. Soft rots cause economic losses by reducing the amount and quality of produce available and by greatly increasing expenses for preventative measures during storage and shipment.

Symptoms

The name soft rot comes from the characteristic soft decay of fleshy tissues associated with the disease. Soft-rot symptoms are similar on all hosts. Carrot roots, celery stalks, onion bulbs, and potato tubers are dormant reserve storage organs high in carbohydrates. On such tissues, soft rot first appears as a small watersoaked spot which enlarges rapidly. The affected tissues become mushy and soft, appear cream colored, and leak water. A whole fruit or tuber may be changed into a soft, watery, rotten mass within 3–5 days. As the tissues collapse, a foul odor is released. Rotten cabbage and onions smell particularly bad. Once you smell a bad case of soft rot, you will never forget it.

Sometimes soft-rot bacteria invade the stems of tobacco plants from which the blooms have been removed (this is done with some cultivars to increase yields). The exposed pith tissue is very susceptible, rots rapidly, and turns brown. The leaf petioles become infected, and the leaves fall off. As the decay spreads down the stalk, it turns brown to black. During wet weather, infected leaves continue to rot in the curing barn, particularly in bulk curers where the leaves are densely packed into bulks and dried (cured) with hot air.

Black leg, the seedling form of soft rot, occurs on tobacco seedlings during wet periods. Leaves that touch the ground become infected, and the rot spreads through the petioles into the stem which turns black. Frequently the disease appears in areas up to 1 m in diameter in seedbeds. The black leg form of the disease can also be a serious problem on potato plants.

Causal Agent and Disease Cycle

Soft rots are caused by *Erwinia carotovora* subsp. *carotovora* (L. R. Jones), Bergey, a non-spore-forming, gram-negative, motile, rod-shaped bacterium (2.0 μm $\times$ 0.8 μm). It occurs singly or in chains; above 27°C it forms longer rods and filaments. Each rod usually has two to six flagella uniformly distributed over the surface. The bacterium can grow aerobically or anaerobically over a wide range of temperatures, but does not grow below 2°C. The optimum growth temperature is about 25°C; the bacterium is killed at 50°C. *Erwinia carotovora* produces pectolytic enzymes which dissolve the cementing substances that hold plant cell walls together. Strains of the bacterium differ in their ability to cause rapid rotting of different hosts.

Soft-rot bacteria overwinter in infected storage organs in plant trash, on seed, in soil, and in the pupae of several insects. Harvest bruises, freeze injury, and insect wounds increase the chances of decay. Abundant moisture at the surface of wounds is necessary for invasion. High RH favors disease development. Larvae of certain maggot flies carry the bacteria; when such insects attack healthy plants or storage organs, they may become infected. After soft-rot bacteria invade plants through wounds, they live and multiply on the liquids released from wounded cells. The bacteria multiply rapidly producing increasing amounts of enzymes that cause the cell walls to soften and dissolve; the contents leak out, and the cells collapse and die. Other contaminating bacteria that are always present help in the decay process. In a few days much of the stored crop can be a rotten, stinking, slimy mass. Instances are known where Irish potatoes harvested, washed, and packaged on Friday and shipped by truck over the weekend were completely rotted when the truck was opened on Monday morning. Moreover, the truck smelled so bad that the driver had trouble getting rid of the mess and cleaning the truck.

Management

Proper handling of crops at harvest is the most important measure for managing soft rot of vegetables. Avoid bruising plant parts and keep plant surfaces dry. If Irish potatoes are washed at harvest before shipment, they should be dried before being sacked. Do not store wet, wounded, or decayed tubers, fruits, or plant parts. Crops should be stored at low RH and at the lowest temperature at which the particular product can be expected to retain its edible quality. Maintain good air flow in storage areas to insure dry plant surfaces. Leafy vegetables should be cooled to 4°–6°C immediately on arrival.

Crops very susceptible to soft rot should be rotated with cereals, grasses, or other nonsusceptible crops. Spraying or dusting to control the insects that spread the pathogen also helps control soft rot.

Bacterial Wilt of Cucumbers

Bacterial wilt is a common and destructive disease on cucumber and muskmelon. Damage to squash and pumpkin is less severe. This disease is unusual in that the pathogen overwinters in the cucumber beetle and could not survive without this insect host.

Symptoms

The first evidence of cucumber wilt usually appears on individual leaves as dull green patches that become limp in sunny weather. As the disease progresses, more leaves wilt; eventually an entire branch becomes permanently wilted with shriveled, dead leaves (Fig. 16.3). Occasionally a sticky ooze can be seen on the fruits. When an infected stem is cut crosswise, the bacterial ooze can be drawn out in strands 2–5 cm or more in length. This is an accurate way to identify the disease.

Fig. 16.3. **Cucumbers infected with bacterial wilt.**
Courtesy Plant Pathology Dept., Cornell University

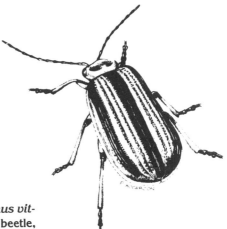

Fig. 16.4. **Adult striped cucumber beetle,** *Acalymus vittata,* **about 5 mm long. The spotted cucumber beetle,** *Diabrotica undecimpunctata,* **also carries the wilt bacterium.**
Courtesy N. C. Agric. Extension Service

Causal Agent and Disease Cycle

The causal agent of cucumber wilt, *Erwinia tracheiphila,* (Smith) Bergey, is a motile, gram-negative rod with four to eight scattered flagella. It forms capsules but no spores. Colonies in agar medium are small, glistening, circular, smooth, and white.

The bacteria overwinter in the bodies of adult cucumber beetles (Fig. 16.4). The only means of natural infection is by beetles feeding upon young leaves or cotyledons in the spring. After infection, the bacteria invade the water-conducting vessels. Here they multiply to enormous numbers; the sticky slime plugs the vessels, preventing flow of sap, and wilting results. Secondary spread occurs only in wounds made by feeding beetles.

Management

Since the bacterium that causes wilt is solely dependent on the cucumber beetle, the most efficient way to manage the disease is to manage the beetle.

Insecticides must be applied early to forestall feeding by overwintering adults and continued as necessary. Some cucumber cultivars are more resistant to wilt than others. These should be used in localities where the disease is a problem.

Bacterial Canker of Tomato

Bacterial canker, although listed as a major disease in several countries, is erratic in occurrence. It generally is not as prevalent as early or late blight on tomato, which are caused by fungi (see Chapter 13). However, when canker does become a problem, there is no disease more serious or devastating. Canker is an important disease only on tomato, but it occurs worldwide in the major tomato-producing areas.

Symptoms

Canker symptoms vary according to environmental factors and age of plant when infected (Fig. 16.5). Seedlings wilt and die. Young plants not yet bearing fruit show a one-sided wilting of the lower leaves. Leaves on older plants die at the edges and dry up, followed by a general withering or firing of the foliage. Yellow to black streaks are found on stems, petioles, and leaf veins. These areas rupture to form brown, pimple-like cankers. Usually the roots are not infected.

During wet weather, secondary spread from cankers is common with infections occurring on the foliage and fruits as watersoaked spots (3mm in diameter) surrounded by a white halo. This condition on the fruit is called *bird's eye* spot and greatly reduces their market value.

Causal Agent and Disease Cycle

Unlike most plant-pathogenic bacteria, *Corynebacterium michiganense* pv. *michiganense* (E.F. Sm.) Jensen, which causes bacterial canker, is a gram-positive rod. It averages 0.6 μm × 1.0 μm, forms yellow colonies on nutrient agar, and has no flagella. Nonpathogenic strains have been reported.

The pathogen enters the host through wounds or natural openings (stomates). All parts of the tomato plant are susceptible including the seeds. Rapid succulent growth of the host, warm weather (28°C), low light intensity, sandy soils, and adequate moisture increase severity of disease. The bacterium survives up to 5 years in soil and 2 years in compost. Scientists agree that the worldwide distribution of *C. michiganense* has resulted from contaminated seed carried from one country to another.

Local spread is possible from diseased transplants, contaminated soil or debris, splashing rain, and by workers handling infected plants.

Management

Bacterial canker can be managed effectively by the use of bacterial-free seed or transplants, crop rotation, and sanitation.

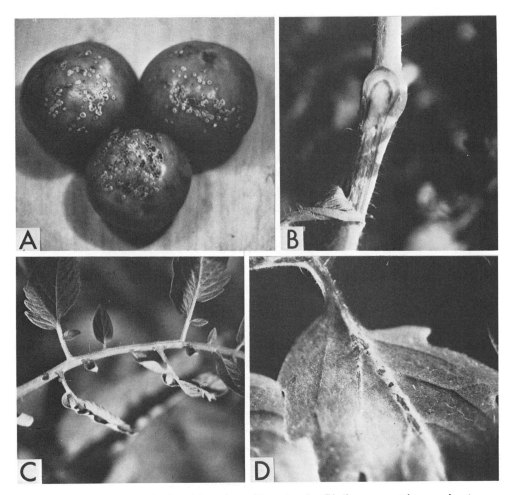

Fig. 16.5. Symptoms of bacterial canker of tomato. A—Bird's-eye spotting on ripe tomatoes. B—Yellow to brown streaks in water-conducting tissue of stems and petioles. C—Wilting of leaflets. D—Cankers along midribs and veins of leaflets.
Courtesy D. L. Strider

Use only certified seed obtained from areas where canker is not present, or treat the seed in hot water (56°C) for 25 min or in 5% hydrochloric acid for 5 hr. Seedbeds should be fumigated or steamed before sowing seed. Roguing, the careful removal of diseased plants, also reduces spread. Antibiotic sprays, especially streptomycin, and copper materials help reduce disease incidence. Since the organism persists in the soil 3–5 years, long rotations are recommended.

Bacterial Spot of Peach and Plum

Bacterial spot appears to be native to North America, where it occurs on wild plum. The disease probably spread from wild plum trees to peach orchards planted nearby. Not only does it weaken the trees, but also it disfigures the fruit. Sometimes in years favorable for the disease, 15–30% of the spotted fruit must be discarded.

Bacterial spot does the most damage on peaches and plums, but almonds, apricots, cherries, and nectarines also are hosts.

Symptoms

Bacterial spot affects leaves, twigs, and fruit. It occurs on the leaves as small circular or irregular areas. The spots gradually enlarge from 1 to 5 mm in diameter, become angular, and turn brown to black. Usually this dead tissue falls out, giving the leaves a ragged, shot-hole appearance. Sometimes several spots may merge and involve large areas of the leaf. Young succulent twigs are killed, and the green shoots turn black, particularly on susceptible cultivars. Small circular brown spots appear on the fruit surface. There may be only a few spots, or they may cover large areas. Later the fruits "pit" or "crack" and become useless for sale.

Causal Agent and Disease Cycle

Bacterial spot of peach is caused by *Xanthomonas campestris* pv. *pruni* (Smith) Dye, a short gram-negative rod (0.5 μm × 1.0 μm), which has one to several flagella attached to the end of the cell. It is aerobic and forms yellow colonies on agar. Exposure to the sun's rays for about 30 min kills the bacterium.

The pathogen overwinters in twig lesions and in the bud scales near ends of the twigs. Primary infection of young leaves and twigs occurs early in the spring. The pathogen can enter uninjured peach tissue through natural openings. Warm weather with light, frequent rains accompanied by strong winds and heavy dew is most favorable for severe infection. When the disease is present, premature leaf fall occurs; the trees are weakened and predisposed to winter injury. If wet weather prevails during the few weeks before harvest, when the fruits are enlarging rapidly, damage is great.

Management

Because bacterial spot is more severe on weak trees that are growing poorly, the first step in managing the disease is to keep trees vigorous. This means trees must be pruned correctly, fertilized properly, and sprayed to manage other diseases and insects. In many areas, especially in light sandy soils, it is hard to keep trees growing vigorously. Therefore, resistant cultivars should be used as recommended by the state experiment station whenever bacterial spot is a problem.

Crown Gall

Crown galls are diseased growths or tumors that appear on apples, grapes, peaches, roses, tobacco, and many other woody and herbaceous plants belonging to 140 genera of more than 60 families. The galls appear commonly on the stems of plants (the crown of roses or brambles) where they come out of the ground. That is why it is called crown gall.

The disease occurs all over the world. Infected nursery stock probably carried the causal agent from one country to another. On some crops, crown gall can cause significant losses; on others, it is just a curiosity.

Crown-gall tumors resemble human and animal cancers. In fact, some scientists call crown gall a "plant cancer." When the cause of crown gall was first discovered about 1900, some researchers thought that by studying crown gall they could unlock the secret of human cancer. Therefore, the causes and mechanisms of plant galls have been studied extensively. The changes that convert a healthy plant cell into one that produces tumorous cells have only recently been found to involve a *plasmid* carried by the causal agent. Much has also been learned about the biochemical processes of growth stimulation and inhibition and the chemical compounds that control them.

Symptoms

Crown gall first appears as small overgrowths or swellings on the stems and roots, particularly near the soil line (Fig. 16.6). Ordinarily, the galls are quite soft. As they enlarge, their surfaces become wrinkled and furrowed, and the tumors turn brown or black. Some tumors are spongy; others become hard and knobby and may reach 30 cm in diameter. Sometimes the tumors rot partially, and new growths appear the next growing season. Several galls may occur on the same root or stem, singly or in groups. Secondary tumors also develop.

In general, tumorous plants are stunted, have small chlorotic leaves, and are more susceptible than healthy plants to adverse conditions, particularly winter injury.

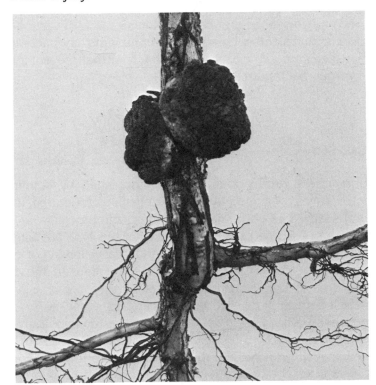

Fig. 16.6. **Crown-gall tumor on peach seedling. Seedlings affected with this disease should not be used for planting stock.**
Courtesy C. N. Clayton

Causal Agent and Disease Cycle

The crown-gall bacterium, *Agrobacterium tumefaciens* (Smith & Townsend) Conn, is a soil-inhabiting, gram-negative rod (0.5 μm × 2.0 μm) with two to four flagella at one end. It forms capsules but no spores. The optimum temperature for growth is 25°–30°C. In some soils the bacterium remains alive for years.

The bacteria enter plants through wounds and occupy the intercellular spaces. Juices from ruptured host cells diffuse several millimeters from the wound site and bathe other cells. Bacterial cells in contact with wounded cells, in some as yet incompletely understood manner, transfer plasmid genetic material to cytoplasmic genetic material of the host. These host cells begin to produce tumor cells within 4–5 days. Growth hormones that control cell enlargement and cell division also are involved in tumor development. Cells in the tumors continue to grow in an abnormal manner until they reach maximum size. Once the cells have been transformed to tumor cells by the bacterial plasmid, the bacterium is not necessary for gall growth.

Management

To effectively manage crown galls, growers should adhere to the following recommendations:

- Use nursery stock free of crown gall.
- Bud instead of graft nursery stock.
- Do not wound crowns or roots while cultivating.
- Rotate infested fields with corn or small grains before planting nursery stock.
- Remove and burn infected trees.
- Do not mix diseased and healthy plants.
- Dip or spray plants with solutions containing antibacterial chemicals.
- Treat seedlings with closely related bacteria that prevent infection by the crown-gall bacterium.

Mycoplasma Diseases

For a long time a group of diseases distinguished by stunting, chlorosis, yellowing of leaves, changing of flowers to leaflike structures, the production of numerous leaves, or combinations of these symptoms were thought to be caused by viruses. Some common diseases of this type are aster yellows, big bud of tomato and tobacco, corn stunt, and lethal yellowing of coconut palm. However, we now know these diseases are not caused by viruses but by organisms, smaller than most bacteria, called *mycoplasmas*.

Japanese scientists, in 1967, were the first to show the presence of mycoplasmas in the cells of infected plants. Mycoplasmas do not have a rigid cell wall, only a cell membrane. These organisms (Fig. 16.7) assume many different shapes; they are soft, fragile, and subject to distortion under slight shifts in osmotic pressure. They reproduce by forming chains, budding, or fission.

Fig. 16.7. Mycoplasma in food-con-
ducting tissue of a tobacco plant. These
microorganisms are soft and fragile.
They are easily distorted by pressure
changes in the host cell.
Courtesy J. F. Worley

Mycoplasmas are transmitted from plant to plant by leafhoppers. An unusual property of mycoplasmas is their ability to multiply in leafhoppers, in some cases with harmful effects such as decreased hatchability of insects from infected eggs and decreased life span of infected adults.

Corn stunt is caused by a spiral mycoplasma or *spiroplasma* transmitted by leafhoppers. It is a serious disease on maize in Mexico. It probably will be a serious problem only in areas with mild winters or where the leafhopper vector can survive because corn-stunt spiroplasma must survive in overwintering leafhoppers. If cold temperatures kill these vectors, the spiroplasmas die, and there is no inoculum left to start a new epidemic.

Mycoplasmal diseases of expensive perennial ornamentals and other valuable plants can be managed or delayed by treatment with injections of certain antibiotics such as tetracyclines. In some cases, management of the leafhopper vector helps delay spread of the disease.

17

Viruses

The discovery of viruses ranks with the conquest of bacteria as another giant step in the never ceasing battle against hunger, famine, disease, and death. We now know that all living things—animals, plants, even fungi and bacteria—are attacked by viruses. More than 400 plant viruses have been described since 1898.

The invention of the light microscope permitted scientists to see and study bacteria and helped produce the germ theory of disease. But until about 1935 many human diseases—such as chicken pox, colds, infantile paralysis, influenza, measles, mumps, smallpox, scarlet fever, and yellow fever—for which no causal organism could be observed, were believed to be caused by "ultramicroscopic organisms" (i.e., invisible bacteria) that were beyond the range of the microscope.

In plant pathology, too, despite careful and extensive research the causes of some plant diseases baffled the best-trained and most efficient workers. Some of these diseases were known to early workers by symptom names (e.g., chlorosis, dwarf, etch, leaf curl, leaf roll, mosaic, mottling, vein banding, and yellows). Many of these puzzling diseases caused a partial yellowing of plants, and some have been shown to be caused by mycoplasma. (See Chapter 15.) Research workers soon learned that some of these diseases were readily transmitted, whereas others were rarely or never transmitted to healthy plants. In the 1890s, scientists determined that tobacco mosaic was caused by a "contagious fluid" that did not contain bacteria.

Just prior to World War II, W. M. Stanley isolated from the leaves of mosaic-infected tobacco plants some tiny white crystals which he claimed were the tobacco mosaic virus. He wrote, "Tobacco mosaic virus is regarded as an autocatalytic protein, which, for the present, may be assumed to require the presence of living cells for multiplication." He was correct, and for this important work Stanley received the Nobel Prize for chemistry in 1946. Also, about this time the electron microscope was invented; scientists soon were able to "see" virus particles, which are so small they are invisible under the highest power of the ordinary light microscope.

Since the 1940s, developments in virology have been rapid and startling. In the early 1960s, F. H. Crick and J. D. Watson working in England set forth their ideas on the double-helix structure of nucleic acids and their replication. As F. C. Bawden had shown, nucleic acids are a vital part of viruses. The work of Crick and Watson permitted scientists to visualize how viruses reproduce themselves and, more than that, showed how genes and chromosomes replicate.

In 1971 T. O. Diener described a new class of infectious units called *viroids* composed only of nucleic acids. These are discussed in Chapter 19.

Naming of Viruses

Ever since viruses were discovered, scientists have argued about how to name them. Even today no one system has been accepted by all. The same virus has often received different names from different workers. Sometimes the same name has been given to distinct viruses. This has occurred largely as a result of naming viruses on the basis of symptoms, without realizing that the same symptoms are not always produced by the same virus or that similar symptoms may be caused by different viruses.

At first, little attempt was made to distinguish between cause and effect. The virus and the disease were usually referred to as the same thing. Later a distinction was made by adding the word virus to the name of the disease to indicate the cause. For example, tobacco mosaic is caused by the tobacco mosaic virus; however, this system of naming viruses on the basis of the host attacked and the symptoms produced is often inadequate. And sometimes it is confusing when names are translated into different languages.

As more has been discovered about viruses, different systems of naming them have been proposed. Viruses can be classified on the basis of several characteristics: morphology (rods or spheres), serology, chemical properties, vectors, hosts infected, or combinations of these.

In this book, virus diseases will be referred to by the name most commonly used. For example, the disease tobacco mosaic (TM) is caused by tobacco mosaic virus (TMV). For convenience, viruses and the diseases they cause are often referred to by their abbreviations consisting of the capitalized initial letters of each word in the name. Thus, maize dwarf mosaic is designated MDM and maize dwarf mosaic virus is designated MDMV.

Properties and Isolation

Viruses are noncellular, ultramicroscopic particles that multiply only in living cells. Most plant viruses consist of protein shells surrounding a core of nucleic acid, which is usually ribonucleic acid (RNA) but in a few cases is deoxyribonucleic acid (DNA). Some viruses are shaped like rods or flexible threads (Fig. 17.1); others are spherical (17.2).

Viruses can be obtained in purified form by grinding up the leaves of infected plants and squeezing the juice through a filter made of glass, paper, asbestos, or other materials that remove bacteria and tissue cells. Then the virus may be precipitated with chemicals, or the plant juice can be placed in a test tube and whirled in a centrifuge at very high speeds to separate viruses from the leaf sap. Viruses withstand freezing, and can remain infective for many months if stored in dry ice or at below zero temperature.

Fig. 17.1. Electron micrograph of potato virus Y magnified about 18,500 times.
Courtesy M. K. Corbett

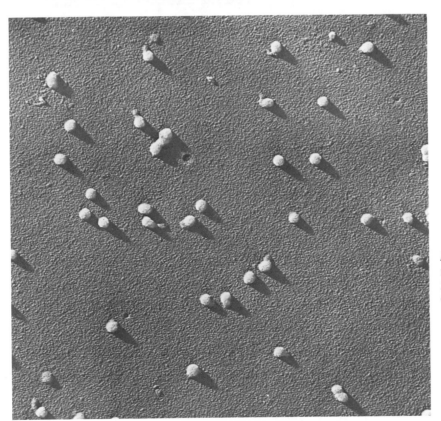

Fig. 17.2. Electron micrograph of tomato spotted wilt virus magnified about 32,400 times.
Courtesy L. M. Black

Growth and Multiplication

Viruses produce no respiratory or synthetic enzymes and are dependent on the host cell for energy and all growth materials. They do not multiply like cells. Outside of, or apart from a living cell, a virus particle can be regarded as a complex molecule of inert organic material. To put it another way, without a living host, a virus is just a complex chemical. But inside a living cell, the virus particles cause great changes in the physiology of the host; they cause disease, sometimes even death, of the host.

Once a virus is introduced into the cytoplasm of a host cell, the protein coat is shed from the nucleic acid and this nucleic acid core "instructs" the cell to begin manufacturing more virus RNA or DNA and virus coat protein. The RNA or DNA and protein are manufactured separately and later assembled in the cytoplasm to form virus particles. All this happens quickly; less than 12 hr after infection, new virus particles can be extracted from the infected cell and adjacent cells. Within a few days the virus can be detected in other leaves; sometimes all growing portions of the plants have virus in them. Plant viruses often multiply within host cells to tremendous numbers. A single cell of a tobacco plant may contain 60 million TMV particles, and the virus may constitute 10% of the dry weight of the plant.

Transmission

Insects, such as aphids and leafhoppers, are the most important agents (*vectors*) for transmitting viruses. Other vectors include mealy bugs, mites, thrips, and white flies. Some plant viruses are also carried by nematodes and soil fungi.

When an insect feeds on leaves of virus-infected plants and sucks out the juice, the insect's stylet or mouth parts become contaminated with virus particles. Then if the insect should happen to fly or crawl to a healthy plant to feed on a leaf, some virus particles are injected into the leaf cell through tiny punctures made by the probing insect or through the damage caused by the insect's chewing.

One important virus—TMV—is transmitted mechanically by contact. Man himself is the principal agent of spread. Once you touch or bruise an infected plant and get a little juice on your fingers and then touch a healthy leaf, you can spread the virus to it. Cultivating machinery also becomes contaminated and can spread the virus up and down the rows of healthy plants. The tobacco in cigarettes, cigars, and snuff may also contain infective TMV. Therefore, if you smoke or chew tobacco while handling plants, you can spread the virus.

Viruses spread easily in greenhouses, seedbeds, or other places where plants are grown close together. It is difficult to keep greenhouse-grown tomatoes free of TMV because the plants are handled repeatedly during transplanting, tieing, pruning, and watering. One infected plant can supply enough virus to contaminate all the other plants.

Management

Virus diseases are hard to manage. Once a plant is infected with virus, it becomes a source of contamination for other plants; therefore, growers should try to produce virus-free plants.

Seedbed soils should be fumigated with methyl bromide or other chemicals that kill weeds (see Chapter 7, Soil Treatment section). Chemical weed management markedly reduces the possibility of virus spread by eliminating hand weeding. A seedbed area should be kept free of virus-susceptible weeds that serve as a virus source.

Because milk inactivates many plant viruses, it is a good idea for workers to dip their hands in milk every 20–30 min when handling plants. Virus contamination also can be removed from the hands by thorough washing with soap and water.

Viruses do most of their damage when they infect young plants. Therefore, crops should be kept virus free as long as possible. This is difficult to do for insect-transmitted viruses, but spraying young plants with insecticides before transplanting can help. Do not grow crops near other host crops, and plant crops during seasons of the year when insect populations are low or before they build up.

Sometimes valuable plants can be freed of viruses by growing the plants at high temperature or by making tiny cuttings or tissue cultures from tips or buds of the growing plants. Often these actively growing bud tissues are virus free; healthy virus-free plants can be propagated from such tissue cultures. This method requires the skill of trained technicians but has been highly successful in the commercial production of virus-free cuttings of carnations, chrysanthemums, geraniums, and other greenhouse crops and some vegetable and field crops. See Chapter 8 (Vegetative Propagation section) for a detailed discussion of this technique.

18
Diseases Caused by Viruses

The symptoms of different virus-caused plant diseases are often quite similar and difficult to distinguish. Thus, identification of a particular disease may require isolation of the causal virus from infected plants and its identification (Fig. 18.1). Various characteristics of isolated virus particles may be used to pinpoint their identity. These include their serological specificity, size and shape as determined by electron microscopy, reactions to different hosts, and chemical composition.

Another approach to identifying viruses, which does not involve their isolation from infected tissues, is *virus indexing*. This technique involves the transfer of a bud, scion, sap, etc., from a plant infected with an unknown virus to one or more kinds of *indicator plants* that are sensitive to known viruses. If the pattern of infection caused by the unknown virus in the indicator plants is the same as that of a known virus, then the two are considered identical.

A few of the more important virus-caused plant diseases are discussed in this chapter. These are summarized in Table 18.1. Readers should be aware that the effectiveness of measures for managing virus diseases often depends upon accurate diagnosis. Although some plant viruses can be identified based on the visible symptoms and general behavior of the diseases they cause, the identity of many viruses can only be determined by trained personnel using a variety of techniques.

Table 18.1. Some Important Diseases Caused by Viruses

Disease	Pathogen	Hosts
Tobacco mosaic	Tobacco mosaic virus (TMV)	Tobacco, tomato, pepper, and others
Vein banding	Potato virus Y (PVY)	Tobacco, potato, pepper
Tobacco vein mottle	Tobacco vein mottle virus (TVMV)	Tobacco
Tobacco etch	Tobacco etch virus (TEV)	Tobacco
Cucumber mosaic	Cucumber mosaic virus (CMV)	Cucumber, bean, tomato, tobacco, and others
Barley yellow dwarf	Barley yellow dwarf virus (BYDV)	Small grains
Maize dwarf mosaic	Maize dwarf mosaic virus (MDMV)	Corn
Maize chlorotic dwarf	Maize chlorotic dwarf virus (MCDV)	Corn

Fig. 18.1.   Isolation of TMV from a hair cell of a tobacco leaf containing a crystal of the virus. A—Leaf before freeze-drying. B—Leaf after freeze-drying. C—Isolated crystal extracted by micromanipulation with a thin needle. × 430. D—TMV particles. × 9,460.
Courtesy R. L. Steere and R. C. Williams

Diseases Caused by Viruses   **255**

Tobacco Mosaic

Tobacco mosaic occurs all over the world. Wherever tobacco is grown, you will find mosaic. Mosaic is common, yet its appearance is sporadic. Epidemics seldom occur, but individual farmers can lose up to 90% of a tomato or tobacco crop.

The annual loss to TMV on tobacco in the United States amounts to 1% of the crop. Losses in other countries are equally great. The biggest losses occur when the plants are infected while young. In such cases both yield and quality are seriously reduced.

Tobacco mosaic is the most important virus disease on tomatoes, especially those grown in greenhouses. Once a greenhouse crop becomes infected, the damage often is so great that the crop is of little value; moreover, it is expensive to clean up the greenhouse to get ready for the next crop. Therefore, every effort should be made to prevent mosaic from getting started.

Much of our knowledge about viruses in general has been derived from research done on TMV. This virus has a wide host range. More than 350 species of plants are susceptible.

Symptoms

Mosaic symptoms vary with the plant species, age of the plant when affected, virus strain, nutrition, temperature and light conditions that occur after infection takes place. Tobacco plants infected with TMV have leaves mottled with light and dark green areas (Fig. 18.2). These areas are most easily seen on younger leaves and produce the mosaic pattern from which the disease gets its name. Some leaves are distorted with dark green blisters or swellings. Stunting of young plants also occurs.

On tomato, mottling occurs on older leaves; the leaflets become long and pointed something like a shoestring. Fruit set is reduced, and those fruits that do form seldom reach full size. Parts of the pulp turn black; fruits from mosaic-infected plants have little market value.

Causal Agent and Disease Cycle

The TMV particle is rod shaped, about 300 μm long $\times$ 15 μm in diameter. The protein subunits are arranged in a helix around a spiral core of ribonucleic acid (Fig. 18.3). The virus is one of the most stable viruses known and can remain infective for at least 50 years in dead, dried tissues and for many months on equipment, flats, greenhouse frames, sawdust, tools, and in the soil.

There are many strains of TMV. Some cause severe damage on a given host, whereas other strains cause such mild symptoms that they are barely visible.

Tobacco mosaic virus is transmitted mechanically. Any means that permit the virus to touch an injured cell of a host plant may result in infection. One of the most common ways the virus is spread is by crop workers with TMV on

Fig. 18.2. Flue-cured tobacco infected with TMV. A, B—Symptoms on young leaves; leaf distortion (B) commonly occurs on plants infected in the four- to six-leaf stage. C, D—Symptoms on old leaves including line patterns along the veins (C) and mosaic burn (D).
Courtesy G. V. Gooding

Fig. 18.3. Model of about $\frac{1}{20}$th of a TMV particle. The protein subunits are arranged in a spiral outside of an inside inner spiral strand of ribonucleic acid.
Courtesy Academic Press, Inc.

their hands. This often happens during weeding, transplanting, cultivating, pruning, suckering, or tying operations. Tractors or equipment that brush against a TMV-infected plant may spread the virus to other plants. Chewing insects such as grasshoppers have been shown to transmit the virus, but this means of spread is not very important.

Tobacco mosaic virus overwinters in a number of ways; these must be understood for successful management through sanitation practices. Infection of tobacco plants from overwintering sources of the virus is known as *primary infection.*

All forms of tobacco, including tobacco products used by humans, may carry TMV. The virus can survive for years in dried tobacco tissue, including trash contaminating field equipment and tobacco sheets used in marketing. The virus also overwinters in stalk and root tissue of mosaic-diseased plants from previous crop. Transplants that come in contact with TMV-infested stalk and root tissue are usually infected.

Such infection occurs during transplanting when a plant is pushed against a piece of virus-infected tissue. The number of transplants that become infected in this way will depend on a number of factors including the amount of mosaic the previous year and the quantity of tissue surviving. The virus can overwinter in dead as well as living plant tissue, but dead tissue contains less active virus than living tissue.

Tomato, pepper, and eggplant also are hosts of TMV and fruit from these crops may contain infective virus. There is, for example, enough active TMV in one infected tomato to infect every tobacco plant grown in North Carolina.

Weeds

A number of weeds are known to be hosts of TMV, including horsenettle (*Solanum carolinense* L.). This weed is common where flue-cured tobacco is grown and frequently is found infected with TMV. The virus can be transmitted from infected horsenettle to tobacco mechanically; this weed (Fig. 18.4) is suspected to be the source of TMV found in many tobacco fields. The only way horsenettle can be eliminated as a source of TMV is to eradicate it by herbicides and/or cultivation.

Fig. 18.4. Shoot from a horsenettle (*Solanum carolinense*) plant. This common weed is frequently infected with TMV and is suspected to be the source of TMV found in many tobacco fields. *Courtesy G. V. Gooding*

Although primary infection can occur anytime during the growing season, the most critical time is during transplanting. Plants infected at this time from any of the overwintering sources will show mosaic symptoms in 3–5 weeks. Generally, if fewer than 100 plants per acre are showing symptoms at this time, secondary spread of the disease can be prevented by removing the diseased plants before cultivation. Plants used to replant rogued plants frequently become infected from virus in the roots left in the soil when plants are removed; so replanting should not be done. If so many plants are infected that roguing is not feasible, secondary spread can be reduced by carrying out cultivation operations so that equipment does not touch the plants. Where this practice is not possible, cultivation should be done when the plants are dry and preferably partially wilted because less virus transmission occurs under these conditions.

Management

The most efficient way to control mosaic is to keep the crop free of TMV. This requires constant attention and an IPM program using every practical management measure available.

Only seed from reputable sources should be used. Tomato seed should have fermented or been treated with acid or bleach by the seedsman. Growers may treat their own seed by soaking it in a 1% water solution of trisodium orthophosphate for 15 min and then in 0.5% sodium hypochlorite (Clorox) for 30 min.

As mosaic infection often begins in seedbeds, growers should produce TMV-free plants and transplant them in noninfested soil. Seedbed soils should be fumigated with methyl bromide or other chemicals that kill weeds and inactivate TMV. Chemical weed management reduces the possibility of TMV spread by eliminating hand weeding. The seedbed area needs to be kept free of susceptible weeds, such as groundcherry or horsenettle, which can carry TMV from one year to the next.

Workers should not smoke or chew tobacco while working in and around seedbeds because their hands become contaminated by contact with tobacco or saliva. This TMV contamination can be removed by thorough washing with soap and water.

Spraying seedbeds with whole or skim milk (2 liters/10 m^2) 24 hr before pulling or handling plants helps to reduce the number of plants that become infected. Thorough coverage is important. Milk sprays should always be used if the use of tobacco by workers cannot be prevented. All workers should also wash their hands with soap or dip them in milk about every 20 min.

Other helpful management practices include removing diseased plants early in the season (do not touch healthy plants during this operation); destroying plants in seedbeds soon after transplanting is completed and in fields soon after harvest; and rotating susceptible crops (tobacco, tomato, and pepper) with nonhost crops (corn, pasture, or grains).

Finally, resistant cultivars should be planted whenever possible. Although mosaic-resistant tobacco cultivars have historically been lower in yield and quality than nonresistant ones, some of the newer releases compare favorably with nonresistant cultivars. Even when resistant cultivars do not perform as well as susceptible ones on a given farm, there are two instances when they would be of benefit. The first is obviously on those farms where losses to mosaic exceed the differences in yield potential between a mosaic-resistant and a susceptible cultivar. Secondly, on farms where monoculture is practiced, a mosaic-resistant cultivar can be grown on a problem field to break the cycle of virus carryover.

Vein Banding, Vein Mottle, and Etch

Potato virus Y (PVY), which causes vein banding, is so named because it was first found on Irish potato. The name is misleading in many areas because the virus may not be a problem on potato, but frequently causes mild to severe diseases on other solanaceous crops, including tobacco and pepper. The virus is probably worldwide in distribution. It is a serious problem on both Irish potato and tobacco in Europe where tobacco and potato production areas overlap. Management of the disease and the insect vector on one crop influences management on the other. Thus, IPM methods for each crop need to be synchronized.

The diseases caused by PVY have not been thoroughly studied in regard to yield losses. Field observations, however, indicate that little loss occurs in tomato and potato. Losses exceeding 50% occur in tobacco cultivars infected with a necrotic strain of the virus, and the severe strain on peppers probably shortens the productive life of these plants. Although not often seen, PVY in combination with potato virus X causes a severe disease of potato known as *rugose mosaic*. Because of the great variation in symptoms caused by different strains of PVY on different crops, field identifications should always be confirmed by clinical tests.

Symptoms

There are a number of strains of PVY based on the symptoms they cause. We currently know of three strains in North Carolina. One causes mild mosaic symptoms on all cultivars of tobacco. One causes vein and midrib *necrosis* (death) on leaves of root-knot-resistant tobacco and mild mosaic symptoms on root-knot-susceptible tobacco. The third strain causes vein and midrib necrosis on both root-knot-resistant and root-knot-susceptible cultivars of tobacco. Ironically, the mild strain on tobacco is severe on pepper; the two necrotic strains on tobacco are mild on pepper. All three strains cause mild mosaic symptoms on potato and tomato.

Causal Agent and Disease Cycle

Potato virus Y is classified with a group of approximately 16 related viruses, including tobacco etch and vein mottle that are long flexuous rods. The individual particles look like short pieces of string (760 nm × 12 nm). The virus is an obligate parasite and cannot reproduce outside a living host.

The only vectors known to transmit PVY are several species of aphids (Fig. 18.5). These sucking insects have mouth parts called *stylets* that are specialized for sucking in a liquid diet. When the aphids insert their stylets into the cells of an infected plant, some of the juice containing PVY particles sticks to the stylets. If the aphid then probes the cells of a healthy plant, the PVY particles are injected into the leaf through the tiny puncture; infection has begun. Aphids can pick up PVY after feeding for 5 seconds and transmit the virus within 10 seconds of feeding.

The perennial weeds ground cherry (*Physalis* sp.) and nightshade (*Solanum aculeatissimum* Jacq.) may serve as overwintering hosts. Potato virus Y is not seed transmitted, nor does it survive in dead host tissue.

Fig. 18.5. The green-peach aphid, *Myzus persicae*, is one of several aphid species that are important vectors for many plant-pathogenic viruses. A—Adult female, about 2 mm long. B—Adult winged male.
Courtesy Nederlandsche Algemeene Keuringsdiest Wageningen, Holland

Management

There are three general approaches to managing PVY: elimination of the virus source, reduction of vectors, and use of resistant or tolerant cultivars.

The PVY sources can be eliminated by plowing under the remains of the preceding crop as soon as possible after harvest so that aphids cannot feed on them. In addition, weed hosts should be removed in fields near Irish potato, pepper, or tobacco. Farmers should grow their own transplants or use only certified transplants or seed. Importation of plants from regions where PVY persists on growing plants the year-round is unwise. Where PVY is found on only a few plants, they may be pulled and removed from the field, thus preventing the virus from being spread to other plants.

The use of insecticides to reduce the number of aphids available for virus transmission helps prevent PVY from getting started in seedbeds. Cultivars resistant or tolerant to PVY should be used wherever possible.

Tobacco Vein Mottle

Tobacco vein mottle is caused by tobacco vein mottle virus (TVMV), which is closely related to PVY. The TVMV is widespread in the Burley tobacco growing areas of the United States. The symptoms of tobacco vein mottle, often confused with vein banding, vary with age of plant when infected and cultivars. The virus is a flexuous rod spread only by aphids; TVMV can be distinguished from PVY serologically. The virus overwinters in perennial weed hosts and living tobacco roots. Management measures are the same as for PVY.

Tobacco Etch

Tobacco etch is caused by tobacco etch virus (TEV), which is another aphid-transmitted flexuous rod related to PVY. Etch occurs in tobacco-growing areas all over the world. It causes serious losses some years on flue-cured tobacco in Canada. There are many strains of TEV which cause symptoms that vary from a mild mosaic or etching to chlorotic tattered leaves and eventual death of the plant. Management measures are the same as for PVY.

Cucumber Mosaic

The virus that causes cucumber mosaic has a wide host range. Many vegetable and ornamental crops are affected both in the greenhouse and field which results in reduced yields of low quality. The disease occurs all over the world. The more important hosts include banana, bean, celery, crucifers, cucumbers, gladiolus, lilies, petunias, and zinnias. Many weeds also carry the virus.

In some areas up to 50% of the plants may be destroyed; susceptible crops such as cucumbers may have to be replaced or abandoned. In Japan where many vegetable crops and tobacco are grown intensively in the same neighborhoods, cucumber mosaic is a constant threat. There, entire communities cooperate in conducting an IPM program for CMV.

Symptoms

Cucumber mosaic symptoms resemble those of tobacco mosaic. Usually first infections on cucumber appear 4–6 weeks after the crop is planted. Young developing leaves become mottled, distorted, and wrinkled. Their edges curl downwards. Growth is reduced, the plants appear stunted, and few runners are produced with only a few flowers and fruit. Older leaves develop chlorotic areas which later kill the entire leaf. Infected cucumber fruits are distorted and misshapen with light and dark green mottling on the surface. Such fruits often have a bitter taste and become soggy when pickled.

On other hosts, symptoms may range from mild to severe depending on the strain and host. Cucumber mosaic on tobacco is hard to distinguish from tobacco mosaic. Tomato leaves become long and stringy like shoestrings. Certain weeds show hardly any symptoms at all.

Causal Agent and Disease Cycle

Cucumber mosaic virus (CMV) is a polyhedral virus whose particles are about 30 nm in diameter. There are numerous strains that vary in pathogenicity. The virus is easily transmitted by rubbing and other mechanical means. At least 60 species of aphids also transmit the virus. Aphids can suck up CMV in less than a minute and inoculate it into plants in less than a minute.

The virus can overwinter in many perennial weeds, flowers, and crop plants such as ground cherry, horsenettle, milk weed, night-shade, ragweed, and various mints. These hosts act as a continuous reservoir of CMV from year to year. The virus remains infective in their roots during the winter. When growth occurs in the spring, the virus is transported to the leaves where it is picked up by feeding aphids and carried to nearby susceptible crops. Once a few cucumbers or squash become infected, the virus is spread during cultivation and harvesting. Sometimes all plants in a field become infected after the first picking. Another source of CMV is infected seed. Several weeds and flowers carry the virus in the seed; however, cucumber does not.

Management

Cucumber mosaic can be controlled, but it takes IPM. Resistant cultivars of cucumber, muskmelon, spinach, and tobacco have been developed. These should be used whenever the disease is a problem. Big losses often can be avoided by preventing or delaying infection in the crop and restricting aphid movement (which is hard to do). Aphids can be reduced by elimination of weed hosts, especially perennials in nearby fields, by disking or plowing under or by spraying the weed source with an aphicide. Aphid transmission of the virus also is reduced by planting the crop during seasons of the year when aphid populations are low or before they build up. Cucumbers should not be grown near other host crops such as tobacco, tomatoes, or peppers. Grow a nonhost crop such as sunflowers between the source of inoculum and the cucumbers to act as a barrier against aphid flights or interplant with barley. Plow under or otherwise destroy the remains of the preceding crop as quickly as possible.

Greenhouse crops should be isolated from other crops susceptible to CMV. Before handling transplants, workers should dip their hands in milk. Insecticides should be used in seedbeds to prevent aphids from infecting transplants.

Virus-free seed should always be used for those crops where CMV is known to be seed transmitted.

The fact that aphids are repelled by certain wavelengths of light and attracted by others offers another possible future method of management, by making plants invisible or unattractive to alighting aphids. Reflective surfaces such as aluminum strips placed between the rows have offered protection in some crops; as the aphids come in to land, the reflected ultraviolet light is thought to act as a repellent.

Barley Yellow Dwarf

Barley yellow dwarf (BYD) is widely distributed throughout the world. It is one of the most important diseases of small grains in the southeastern United States. It affects barley, oats, rye, and wheat; oats are the most seriously affected and suffer serious losses annually. The barley yellow dwarf virus (BYDV) also attacks many weed and pasture grasses.

The yield loss from BYD depends on the time of infection. Plants infected when young may die or if they live, produce no grain; however, infection at heading time may reduce yields only slightly. Infection by BYDV also makes plants more susceptible to winter killing and to infection by root-rotting and leaf-spotting fungi.

Symptoms

Symptoms of BYD include yellowing and dwarfing of leaves, stunting of the plant, reduced tillering, sterility, and shrunken kernels; root systems are also reduced.

The leaves of infected oat plants usually turn red or bronze. The disease on oats has been called *red leaf*. The red color appears to be the result of reduced amounts of chlorophyll and increased amounts of various yellow and red pigments. In wheat, barley, and rye, infected leaves usually turn yellow beginning at the tip and progressing toward the base. Leaf veins frequently remain green longer than the area between the veins and form green lines that project into the yellow area. The disease usually occurs in patches in the field and is frequently more prevalent along the edges of the fields.

Causal Agent and Disease Cycle

Barley yellow dwarf virus is a polyhedral particle about 30 nm in diameter. The virus concentration in infected plants is very low. It is not mechanically transmissible and is not seedborne. Barley yellow dwarf is difficult to diagnose because at present there is no satisfactory method of routinely detecting the presence of the virus in plants. The virus cannot survive from season to season in the soil, nor in dead straw or debris; it survives only in living plants or in living aphids. The spread of BYDV depends upon the spread of the insect vectors. At least nine species of aphids can transmit BYDV.

Some aphid species transmit BYDV more efficiently than others. There are a number of strains of the virus; certain strains can be transmitted by some species of aphids, but not by others. The most important transmitter of BYDV in the fall in the southeastern United States is the oat-bird cherry aphid, a small brown aphid that usually feeds on the lower parts of the plants. In the spring the most common transmitter is the English grain aphid, a larger green aphid that usually feeds on the lower portions of the plants.

The virus remains infective over winter in bromegrass, orchard grass, ryegrass, tall fescue, and other grasses that harbor the aphids. The aphids then crawl or fly from the infected grass to grain crops and carry the virus from grass to grain. The only way by which small grain plants can become infected with BYDV is from feeding by aphids carrying the virus.

When winters are mild, there is abundant survival of aphids in the southern United States. Warm south winds blow or carry the aphids northward in stages as spring advances. The aphids feed on the new tender leaves of wheat, oats, and grass hosts as they begin to grow. Mean daily temperatures of $10°-18°C$ are most favorable for succulent grass growth, aphid activity, and reproduction. Thus, epidemics of barley yellow dwarf do most damage when the spring and early summer weather is cool and moist.

Symptoms usually appear 3–6 weeks after infective aphids feed on plants. The first leaves to show symptoms are best for diagnosis.

Management

As is the case with other viruses transmitted by aphids, management of BYD depends on managing the aphid vector. Attempts to control aphids by spraying in the field have not been satisfactory. Infective aphids that move into a field can transmit the virus before the insecticide kills them.

Several cultural practices help reduce losses. Do not plant small grains near large grass areas, such as pastures, that may act as a source of virus. Do plant oats or barley near the end of the normal seeding period as BYD is generally more prevalent in early seeded grain.

All small grain cultivars are subject to damage by BYD, but some are more affected than others. Therefore, growers should use only those cultivars recommended for their area.

Maize Dwarf Mosaic and Maize Chlorotic Dwarf

Two major virus diseases of corn are maize dwarf mosaic (MDM) and maize chlorotic dwarf (MCD). Before 1960, virus diseases of corn did not receive much attention or were overlooked or confused with other growth problems. However, we now know that these two diseases can cause losses of 5–90% in some fields. Maize dwarf mosaic is worldwide in distribution. Less is known about MCD; however, corn scientists are on the lookout for MCD and reports from new areas are being made.

All types of corn—field, pop, and sweet—are hosts of the viruses that cause MDM and MCD. Johnson grass is a principal host; other weeds including barnyard grass, crabgrass, foxtail, and quack grass also are hosts. Pearl millet, sorghum, Sudan grass, sugarcane, and teosinte also serve as hosts.

Symptoms

Many virus diseases are not easily diagnosed, as the symptoms are similar to those caused by other growth problems. Therefore, it is necessary to be sure that other factors including nutrition, herbicides, soil, or other pathogens are not causing the trouble.

Symptoms of MDM and MCD overlap and may be influenced by environmental conditions and the cultivar of corn grown.

Maize dwarf mosaic first produces a faint, pale green streaking, mottle, or mosaic in younger leaves along the veins. Spots of dark green tissue appear on a lighter green background. As infected plants mature, this discoloration increases and the leaves become yellowish-green. Infected plants may be stunted, bunchy, tiller excessively, produce multiple ear shoots, but set few seed. Early infection may predispose the corn plant to root and stalk rots and sometimes premature death.

Maize chlorotic dwarf, which used to be called *corn stunt*, causes more severe stunting than MDMV. Infected leaves turn yellow, but no mosaic pattern develops. As the plants mature, the yellow discoloration increases; the leaves become red. Infection with MCDV stunts or shortens plants giving them a "bunchy" appearance. Infection often occurs late in the season. On such plants the lower portion grows normally with the upper portion red and stunted. If susceptible cultivars are infected early in the growing season, they produce only short, stunted ears (sometimes called "nubbins").

Causal Agents and Disease Cycle

There are several strains of MDMV, which belongs to the PVY group. The virus particles are flexuous rods, about 750 nm $\times$ 12 nm. Particles of MCDV are small spheres that average about 25 nm in diameter; they can persist in the leafhopper vector for several months. Both viruses are transmitted by insects: MDMV by aphids and MCDV by leafhoppers. Soil and seed spread have not been shown. Johnson grass is the most important overwintering host. Aphids and leafhoppers feed on infected Johnson grass and carry the viruses to corn.

Management

Management of MDM and MCD is based primarily on the use of resistant cultivars and elimination of Johnson grass, the principal host. If these diseases have been identified in an area or have caused losses on a farm, resistant or tolerant cultivars should be grown. Resistant corn cultivars adapted to particular areas are generally available. The county extension agent has the latest information on hybrids suitable for a local area.

Particular attention must be paid to controlling Johnson grass—in the field, fence rows, road sides, and other areas adjacent to corn fields. An efficient weed management program is essential for control of MDM and MCD.

A good corn production program must be followed. Prepare the land properly, use proper amounts of lime and fertilizer, plant corn at the correct time and depth, rotate it with other crops, and use herbicides if necessary.

As corn yields gradually creep upward, it may become feasible to use systemic insecticides on corn crops. Compounds that act both as nematicides and insecticides are now being applied to corn soils before planting. An unexpected benefit of such compounds may be the help they give in controlling the vectors of MDMV and MCDV.

Viroid Diseases

Viroids are one of the newest groups of pathogens described by scientists. Viroids are small units of ribonucleic acid (RNA) that can infect plant cells, replicate in these cells, and cause disease. The RNA in viroids has a much lower molecular weight than the RNA in most plant viruses, and viroid RNA, unlike that of plant viruses, is not enclosed in a protein shell. The way in which viroids multiply is not well understood.

At least five diseases are known to be caused by viroids: potato spindle tuber, citrus exocortis, chrysanthemum chlorotic mottle, chrysanthemum stunt, and cucumber pale fruit. Cadang-cadang disease of coconut palms may also be caused by a viroid.

The symptoms caused by viroids in plants vary and are similar to those caused by virus infection. Symptoms of viroid diseases include dwarfed or stunted plants, rolled or twisted leaves, and mottled or yellowed leaves.

Viroids are spread from diseased plants to healthy plants primarily by mechanical means. Possible ways in which viroids can be spread are in plant sap carried on workers' hands and tools during propagation and other cultural practices and in propagation material such as cuttings or tubers.

Viroid diseases can be managed by using viroid-free propagating materials, by eradicating viroid-infected plants, and by sterilizing tools to remove viroid-infected plant sap before they are used with healthy plants. Workers should also wash their hands thoroughly after handling infected plants.

19

Diseases Caused by Parasitic Plants

Usually we think of parasites as lower forms of plant life such as bacteria, fungi or nematodes. However, a number of flowering or seed plants are parasites on other plants. Parasitic seed plants vary widely in their dependence on a host. The more independent ones, called half-parasites or hemiparasites (e.g., orchids), have chlorophyll and roots and can manufacture their own food; but they depend on their host for dissolved minerals and organic substances. Others, such as mistletoe, have chlorophyll but no roots and depend on their host for minerals and water. Some other parasitic seed plants, having neither chlorophyll nor true roots, depend entirely on their host for their existence.

Parasitic seed plants cause disease principally by depriving their hosts of water and nutrients and by disturbing the balance of growth regulators. The full spectrum of diseases caused by parasitic seed plants is still not known. Some of the important parasitic seed plants are discussed in this chapter. They are summarized in Table 19.1.

Table 19.1. Diseases Caused by Parasitic Plants

Parasitic plants		Host(s)
Common name	Scientific name	
Broomrape	*Orobanche* spp.	Lettuce, tobacco, tomato, other vegetables, weeds
Witchweed	*Striga* spp.	Corn
Leafy mistletoe	*Phoradendron* spp.	Hardwood trees
Dwarf mistletoe	Arceuthobium spp.	Coniferous trees
Dodder	*Cuscuta* spp.	Many crops

Broomrapes

The broomrapes are leafless herbs which appear above ground as clumps of brown, purple, white or yellow stems. Most broomrapes belong to the genus *Orobanche.* They are widely distributed.

Broomrape attacks lettuce, tobacco, tomatoes, and other vegetables and may live on weeds between crop plantings. The broomrape plant arises from the ground at or near the base of the host plant (Fig. 19.1). The stems are 15–50 cm high. Exudates from the growing roots of a host plant stimulate broomrape seeds to germinate. The slender root of the germinating seed attaches itself to the root of the host, forms a tuberous enlargement, and absorbs food from the host. As a result, the weakened host is stunted, does not grow properly, and has reduced yields.

A mature broomrape plant may produce up to one million dustlike seeds, which are easily spread by soil, water, and wind. They may remain dormant in the soil for years.

Fig. 19.1. **Four broomrape plants attached to roots of a tomato plant.**
Courtesy Calif. Dept. Agric.

Management

Broomrape is hard to manage. Hand weeding is often too expensive, the seed lives too long for rotations to be practical, and resistant cultivars have not been developed. Methyl bromide can be used to kill seeds in infested soil. In some developing countries, geese are allowed to graze in infested fields. Two geese per hectare grazing on growing broomrape shoots are said to give good management.

The most promising management methods involve injecting the soil with chemicals that stimulate the seeds to germinate in the absence of the host, treating the soil with herbicides that kill the broomrape and not the host, or spraying the broomrape with a selective herbicide after it emerges from the ground.

Witchweed

Witchweed is a pretty plant with bright red flowers. It grows in clumps to about 50 cm in height. It looks innocent enough but underground it is attached to the roots of the host plant which it robs of food and nutrients.

Witchweed is related to broomrape and belongs to the genus *Striga*. It is a serious pest of corn, sorghum, and sugarcane; it also grows on crabgrass. Witchweed was discovered for the first time in the United States in 1956 in North Carolina. As a result of effective quarantines, the parasite appears to have been limited to the area of original infestation. However, a constant watch is maintained for outbreaks in other states.

Plants attacked by witchweed look like they are suffering from drought. They are stunted, wilted, and yellowish. Heavily parasitized plants eventually die and produce little or no yield.

A single witchweed plant may produce half a million tiny brown seeds, which are easily spread by water and wind. Most seeds require a rest period before they will germinate. Chemicals that leak out of the host roots stimulate the seeds to germinate. One of these stimulants is ethylene. Each seed produces a tiny rootlet that grows toward a host root, attaches itself, and draws sap from the host for its own nourishment. The witchweed plant then produces more rootlets which attach themselves to other roots of the host. Sometimes several hundred witchweed plants may parasitize a single corn plant at the same time. Such plants often die prematurely, especially during dry weather.

It takes about 30 days after the seed germinates for the witchweed plant to emerge above ground, and another 30 days before it flowers. Seeds are produced within 5–6 months. Although the witchweed plant has leaves containing chlorophyll and can probably manufacture its own food, it depends upon the host for water, minerals, and organic substances as well.

Management

Several practices help to manage witchweed. Movement of seed to new areas on transplants, machinery, soil, or agricultural products should be prevented. Injecting soil with ethylene gas induces witchweed seed to germinate in the absence of a host plant. Ethylene can also be injected during the growing season in fields growing such nonhosts as cotton, soybeans, or tobacco.

Herbicides also help. They may be used in corn fields before witchweed emerges; or specific chemicals may be used after emergence. Seedbeds can also be treated with methyl bromide.

Mistletoes

Three hundred years before Christ, the Greek botanist Theophrastus wrote down observations that showed he recognized mistletoe as a parasitic plant. Mistletoe has been featured in European legend for centuries. It was considered to have unusual medical and magical value and was frequently used in religious rites. The early American settlers recognized mistletoes in the hardwood forests of the eastern United States and perpetuated the legends and symbolism that had developed in their homelands. Today, mistletoe is highly prized for Christmas decorations. It is often hung in doorways during the holiday period. Those who stand beneath it may receive an unexpected kiss or some other blessing.

There are two kinds of mistletoe, leafy and dwarf. In most parts of the United States the leafy mistletoes are scarce and regarded more as botanical curiosities than as serious pests. Most of the mistletoe for the Christmas trade is gathered in the southern United States where it is abundant in some localities and provides off-season income for farm workers. Most leafy mistletoes found in the United States belong to the genus *Phoradendron*. They grow principally on hardwoods but also parasitize cedar, citrus, cypress, juniper, and pecans. *Phoradendron* spp. cannot stand cold winters and seldom occur in the forests of the northern United States (Fig. 19.2).

The dwarf mistletoes are a serious pest in pine forests of the western United States. Trees of any age may be dwarfed, deformed, or killed. Timber quality is reduced and affected trees produce less seed and are subject to wind breakage, insects, and other diseases. Seedlings and young trees are often killed by dwarf-mistletoe infections.

Infected branches are easily recognized by the branched shoots of the mistletoe plants which occur in tufts or distributed along the twigs of the host.

Fig. 19.2. Clusters of mistletoe plants growing on an oak tree.
Courtesy L. F. Grand

Diseased limbs become swollen and branch excessively forming witches' brooms. Heavily infected stands contain deformed, stunted, and dead trees or trees broken off at trunk cankers.

Dwarf mistletoes that parasitize conifers belong to the genus *Arceuthobium;* all species are regarded as hemiparasites. They produce clusters of green, perennial, jointed stems, up to 10 cm long, on the branches and trunks of the host. Usually the inconspicuous leaves are green, leathery, and scalelike. The stems are supplied with water and nutrients from a rootlike absorbing system, made up of haustoria, which developed in the bark and wood of the host. The flowers are either male or female. Pollinated female flowers produce shiny, white explosive berries in late fall or early winter. Within each berry is one seed coated with a sticky substance called *viscin.* When the fruits are ripe, they develop an internal pressure. If fruits are touched or disturbed, the ripe seed may be shot 12 m or more away. The sticky seeds adhere to anything they touch. Birds also feed on the pulp of the berries and carry sticky seeds from one place to another. The seeds may germinate within a few weeks or delay until the next spring.

A mistletoe plant removes water and nutrients from its host and saps the vitality of the entire tree. It also upsets the hormonal balance with resultant swellings and deformities of the limbs and excessive branching.

Management

At the present time, clear cutting (the practice of cutting every tree of a given kind or species in a selected area) is one method of reducing losses caused by dwarf mistletoe. It is followed by burning to eliminate any undetected infection. In the past, forest fires helped keep mistletoe in check, but with improved fire control mistletoe continues to spread. For large infection centers where burning all infected trees is not feasible, it is necessary to plant a nonsusceptible host between infected trees and newly cleared areas. In areas where dwarf mistletoe is a problem, trees should not be cut when berries are ripe and seed dispersal in progress. Pruning of affected branches may be effective in high-value stands if infection is light.

Chemical management of dwarf mistletoe with herbicides is being investigated. Biological management using fungi that attack the mistletoe or insects that feed on it also show some promise. Resistant cultivars are not available.

Dodder

Among the unique types of seed plants that parasitize green plants are the dodders, members of the genus *Cuscuta.* There are more than 100 species which grow almost everywhere within the temperate and tropical zones. Dodder plants appear as conspicuous tangles of intertwined yellow threads on the aerial parts of its host plants. Sometimes dodder is tinged with red or purple. Occasionally it is almost white. The plant has a number of unusual and poetical names such as devil's hair, gold thread, hair weed, hell vine, love vine, pull down, and strangle weed.

Dodder is particularly troublesome in regions where clover and alfalfa are grown extensively. Dodder reduces growth and yield of infected plants and clogs up harvesting machinery. In some European countries production of clover seed is difficult because of dodder. Other crops that suffer losses are flax, lespedeza, onions, potatoes, sugar beets, and several ornamentals. Cereal plants are never attacked.

Dodder seed are gray to red-brown and somewhat smaller than alfalfa seed. The seeds overwinter as contaminants among crop seeds or in the soil where they may survive for 5 years. When a seed germinates, it produces only a rudimentary root system and sends up a yellow shoot whose tip grows in a spiral fashion (Fig. 19.3). If the shoot touches the stem of a susceptible plant, it twines around the stem and quickly produces haustoria which penetrate the stem and absorb juices from the plant. The dodder continues to grow and may cover many nearby plants with its colorful net of threadlike stems. If the dodder shoot finds no host, it may live for several weeks and then die of starvation.

Management

Using dodder-free seed and transplants and fumigating seedbed soils are important in managing dodder. So is thorough cleaning of equipment before moving it from dodder-infested to dodder-free soils. Animals should not be grazed in dodder fields and then moved to dodder-free soils, as the seed is spread in the manure. Spot infestations can be sprayed with contact herbicides. Heavily infested fields can be treated with soil herbicides.

Fig. 19.3. Growth of dodder on a host plant. A—Sprouting seed with tendril-like filament. B—12 hr later, the filament has extended upward and outward. C—24 hr later, the dodder has twined itself around the host's stem. D—Close-up of dodder on chrysanthemum showing coils and haustoria.
Courtesy R. E. Hutchins

20

Abiotic Agents

Some plant diseases are not caused by living agents. Instead, they are caused by an upset in the environment—usually too much or too little of something necessary for growth—or by the presence of some harmful substance in contact with the plant. We call these *abiotic diseases* or *noninfectious diseases*. Noninfectious diseases occur in the absence of pathogens and, therefore, cannot be transmitted from one plant to another. These diseases can occur at any stage from seed to mature plant; they may cause damage in the field, during storage or transit, or at the market. Symptoms may range from slight to severe, and affected plants may be killed.

Plants, like every other living thing, require specific environmental conditions to prosper. Some plants thrive under moist conditions, others do well in dry atmospheres; some plants grow fastest in bright sunshine, others need shade; some require fertile soils, others do not. Whenever the proper balance of temperature, soil moisture, light, purity of air, nutrients, and pH is upset to a smaller or larger degree, this imbalance is reflected in the growth of plants. Moreover, cultivated plants are frequently grown in artificial environments or subjected to cultural practices (e.g., fertilization, irrigation, pesticide or hormone treatments) that may affect their growth considerably.

Deficiencies and Toxicities

Nutrient Deficiencies

Plants require mineral elements for normal growth. Nitrogen, phosphorus, potassium, calcium, magnesium, and sulfur are needed in large amounts and are called *major elements*. Iron, boron, manganese, zinc, copper, molybdenum, and chlorine, needed in very small amounts, are called *trace* or *minor elements*. When one or more of these nutrients are lacking or deficient, plants become diseased and exhibit various symptoms that may appear on all organs, including flowers, fruit, leaves, roots, seed, and stems.

Certain symptoms are the same regardless of the missing element, but other diagnostic features usually accompany a deficiency of a particular nutrient. The lack of normal amounts of most essential elements usually results in a reduction of growth and yield. As the deficiency becomes greater, symptoms become more pronounced, and affected plants may even die. Characteristic symptoms of specific deficiencies include the following: nitrogen-deficient plants grow poorly and are light green in color; calcium-deficient plants have young leaves that are distorted with hooked tips and curled margins; boron-deficient plants often show brown or dead patches of tissue in roots or fleshy organs.

Nutrient deficiencies can be corrected by fertilization with the missing elements. When a deficiency is suspected, samples of soil and plant tissues should be sent to a soil analysis laboratory, which can determine what elements are lacking. Based on the test results, growers can devise fertilization programs to supply the missing nutrients.

Nutrient Toxicities

One of the biggest problems today of the home gardener or owner of house plants is overfertilization (Fig. 20.1). Fertilizer, when applied in excessive amounts or placed incorrectly, will "burn" roots. Root burning occurs when the concentration of salts or fertilizer in the soil water becomes so high that water is drawn from inside the roots (*reverse osmosis*); they turn brown (burn) and die. Evaporation can concentrate salts in the root zone and also burn roots. This often happens with a potted plant. Table salt applied to icy roads in winter can damage roadside plants the following spring when the dissolved salt washes off the road.

In addition, there should be a proper balance of nutrients in the soil. If certain nutrients are in excess, they may interfere with the uptake of others.

Soil analyses help determine if certain elements are present in excess. Also, the level of soluble salts can be determined, which indicates if too much fertilizer has been applied. Homeowners should be careful not to overfertilize potted plants, shrubbery, or lawns.

Fig. 20.1. This poinsettia plant is dying from too much fertilizer. The high level of soluble salts in the soil causes water to be drawn from the plant roots (reverse osmosis) and the plant dies.
Courtesy N. C. Agric. Extension Service.

Acid and Alkali Soils

Most plants will not grow in extremely acid (sour) or extremely alkaline (sweet) soils. The acidity or alkalinity of soil is measured by a scale of pH units on which 7 is neutral, 0 is the most acid, and 14 is the most alkaline. Most soils range between 4 and 8 on the pH scale. In acid soils, mineral salts become more soluble and may become so concentrated that they are toxic to plants or interfere with the absorption of other necessary elements. Symptoms of mineral deficiency then develop.

Excessive amounts of certain salts raise the soil pH and cause alkali injury. This often happens when desert soils are irrigated.

It is relatively easy to determine the pH of soil with instruments called pH meters. Most soil laboratories have them. All farmers and gardeners should know the pH of their soil. Once the pH is known, steps can be taken to change the pH to a range favorable to plant growth. Acid soils can be corrected by spreading lime on the land.

Weather Extremes

Frost and Cold

Low temperatures injure plants primarily by causing ice formation, between and/or within cells, that injures membranes and other cell components. Low temperatures, even if above freezing, may damage warm-weather crops such as corn and beans. Plants suffer a variety of injuries from temperatures below freezing. Late frosts kill tender plants, buds, leaves, or flowers. Fleshy tissues such as potato tubers become black and blotchy. Low winter temperature may kill roots of trees and also cause bark-splitting and cankers.

Rapid decreases in temperature cause more damage than slow decreases. Rapid increases in thawing have the same effect. Frost-injured plants wilt quickly as soon as the sun hits them, and in a few days they turn brown.

Valuable plants can be protected from extremes of temperature by planting them at the proper time. Proper shading, covering, mulching, irrigating, fertilization and heating (during cold snaps) will also help.

Heat

Leaf scald ("sunburn" or "sun scald") occurs during hot weather after a plant has made rapid growth, especially after periods of rainy, cloudy weather. Large, irregular, watersoaked or dead areas may form on the sun-exposed sides of fleshy fruits and vegetables, such as apples, tomatoes, onions, and potatoes. Succulent leaves may also be injured.

Hot-weather injury is increased by the excessive light, drought, high winds, and low RH that usually accompany high temperatures.

Lightning

Lightning frequently strikes in cultivated fields. Growers often mistake lightning injury for an outbreak of some new and destructive disease, because the area has the appearance of a spot in which a disease is spreading (Fig. 20.2). The affected area in a field is usually circular and may be 15 m or more in diameter. The plants in the center of the area may be killed to the ground line, the injury being less severe toward the borders. Roots of injured plants show a broad brown streak from the soil level upward along one side of the stem into the midribs of the leaves. The injured midribs and veins on the lower leaf surface turn black and collapse. Some leaves may be brown and dry at the tips along the midveins. The uneven growth of the injured midrib and the normal leaf blade results in leaves becoming puckered and distorted. Also the continued growth of the uninjured side of the whole plant and the simultaneous failure of the injured side to elongate may cause the stem tip to turn and twist. The leaves may be green for several days after the stalk and midribs are shriveled and dead. When the stalk of an affected plant is split open lengthwise, the pith is found to be separated into disks, which gives a peculiar ladder-like effect.

Fig. 20.2. **Lightning injury on cotton may cover an area 15 m in diameter. Plants near the point of strike usually wilt and die; plant damage decreases near the outer edge of the strike zone. Symptoms can be confused easily with high application rates or carryover of herbicides.** *Courtesy N. C. Agric. Extension Service*

Abnormal Water Conditions

All living things are composed mostly of water—often 80–95% by weight—and require water to live. It acts as a solvent for and transports food and food materials from soil into a plant and from cell to cell throughout a plant. It serves as one of the crude materials in the formation of sugars by photosynthesis. It preserves turgidity within cells. It permits transpiration and thus promotes and regulates growth.

Plants are very sensitive to moisture. Too much water or too little will cause injury. Water must be available at the correct time in the correct amounts. The health of a plant is affected by the moisture supply of the soil and air. The amount of water needed for proper growth is influenced by temperature, sunshine, soil structure, and wind.

Drought

The first symptom of water shortage in plants is wilting, which results in drooping of succulent shoots and rolling of the leaves. This is a common sight on a hot, summer day. Plants may recover during the night, but if the lack of water continues, they become dwarfed and stunted. The leaves yellow or redden, begin to die along the margins and tips, and finally drop off. Without sufficient water, root crops remain small; cereals produce shrivelled grains; fruits become spotted, deformed, or stunted and may shrivel and fall prematurely. Prolonged drought leads to plant death. Drought also may increase the damage from certain root diseases.

Obviously, the solution to a shortage of water is to supply more of it. Nature, however, does not always cooperate; farmers gamble each year that there will be enough rain at the right times. In other areas, irrigation is widely used to supply water at the proper time. However, irrigation is expensive and can be used profitably only on high-value crops.

Sometimes house plants, greenhouse plants or seedbed crops are neglected, and the plants become dry and wilted. If so, they should be adequately watered as soon as possible.

Drowning

Most cultivated plants cannot obtain enough oxygen from water; hence, a soil saturated with water will not have enough oxygen for the root system—and oxygen is necessary for life. When soils are flooded, poorly drained, or waterlogged, many plants that cannot stand "wet feet" will grow poorly, drown, or die.

However, lack of oxygen is not the entire explanation of injuries due to too much water. Roots injured from oxygen deficiency are often attacked by bacteria. The dying root cells and bacteria produce toxins that injure the plant and prevent proper absorption of water; the plants dry out and the leaves wilt and droop. Injury is more rapid at higher temperatures probably because roots require more oxygen and bacteria grow faster at higher temperatures. All degrees of injury may result: destruction of the entire root system and complete wilting to death of only a few of the deeper roots, in which case only temporary wilting occurs.

Drowning can be prevented by good drainage; thus, management of erosion and drowning are closely connected. For row crops, the rows should be located to decrease runoff and soil loss, low areas should be tiled or drained, and terraces should be maintained properly with adequate ditches provided.

Seeds sown too deeply or plants set too deeply in wet soil often suffer from drowning injury. Adequate drainage should always be provided.

Homeowners should avoid overwatering and possibly drowning their plants.

Phytotoxic Chemicals

Pesticides

Since 1940, herbicides have come into wide use. These chemicals are used to kill weeds in many different crops. Herbicides are highly specialized: chemicals suitable for one type of crop may not be applied to another. The rapid introduction and increasing use of chemical weed management in crop production has been one reason for dramatic increases in crop yields. It has also resulted in wide misuse and carelessness by some growers and homeowners. Plant injury (Fig. 20.3) then may occur.

Occasionally, the misuse of insecticides and fungicides also results in plant injury. Many insecticides and fungicides are now mixed together in the same tank mix and applied simultaneously. This saves time and money. However, some pesticides are incompatible; that is, when they are mixed, reactions occur that reduce their effectiveness and also may cause injury to plants. Often, the injury may be caused in part by the diluents, solvents, or carriers used to dilute the insecticide or fungicide.

Fig. 20.3. **Excessive rates of nitralin applied the previous year along the edge of this field caused severe stunting and purpling of the corn plants, which is similar to phosphorus deficiency. Injured plants have a greatly reduced root system that cannot absorb enough phosphorus to supply the plant's needs.**
Courtesy N. C. Agric. Extension Service

Symptoms of Pesticide Injury

Symptoms of herbicide injury vary widely (Figs. 20.4–20.6). Malformation, stunting, yellowing and/or death of leaves and shoots quickly develop, often overnight or in a few hours. Although herbicide injury may be confused with virus symptoms, it is usually possible to determine whether herbicides might have been used nearby and caused the damage.

Fig. 20.4. Malformation of the cantaloupe on the right was caused by an over-the-top postemergence application of DCPA.
Courtesy N. C. Agric. Extension Service

Fig. 20.5. Veinal yellowing of grape leaves can be caused by an overdose of diuron. This injury usually occurs during a wet season.
Courtesy N. C. Agric. Extension Service

Fig. 20.6. Spray drift or vapor drift from the growth regulator herbicide 2,4-D caused this injury to tomato leaflets. Notice the curved feather-like leaves.
Courtesy N. C. Agric. Extension Service

Other types of injury include dead, burned, or scorched spots on or at the tip of leaves; russeting of fruit; misshapen fruit, leaves, or plants; off-color and/or yellowing or delayed development; poor germination; and complete death of plants. Unfortunately, these symptoms of chemical injury often do not appear until several days after exposure to a herbicide. If they do appear, they are not always clear-cut and often may be confused with other problems. It may be necessary to call on experienced individuals to help identify the cause of the symptoms.

Symptoms of insecticide and fungicide injury to plants are generally similar to those of herbicide injury. They include stunting, yellowing, leaf burn, malformation, and death of certain tissues. The degree of injury is often influenced by temperature and RH.

Management of Pesticide Injury

There are three principal causes for chemical injury to crop plants: misuse and misapplication of pesticides, drifting of pesticide vapors beyond the target area, and pesticide contamination of soil and water. The most effective management of chemical injury is to avoid it by applying appropriate pesticides at the proper rates and in such a way as to reduce drift and contamination of soil and water. See Chapter 7 for a detailed discussion about the proper use of pesticides and the problems associated with their misuse.

Smog, Smoke and Air Pollution

As the population increases and energy demands increase and the number of cars and smokestacks increase, smog (smoke + fog) naturally increases. The Los Angeles-type smog contains oxides of nitrogen, ozone, and other gases that are formed when exhaust fumes from cars or buses react in the presence of moisture and sunlight. The London-type smoke results from the burning of coal, which releases sulfur and other types of gases.

Smog can be blown many kilometers and injure the tender tissue of rapidly growing plants by causing leaf burning, leaf spots, silvering, specking, flecking, or rusting. Trees, flowers, and many other plants are very susceptible to smog. Affected plants age early, and the foliage becomes discolored (Fig. 20.7). Sometimes they are killed.

Air pollution is a community problem, particularly in large cities. As long as we have too many automobiles and smokestacks in big cities producing too much smog, or other toxic substances, plants will become injured and die prematurely. The only sensible management seems to be to reduce the amount of poisonous gases in the air. This feat will be hard to do. It will take community action—and action by nations. Meanwhile, people and plants will suffer. If the air is too dirty for plants, it is too dirty for us to breathe.

Acid Rain

When pollutants in the air such as sulfur and nitrogen oxide combine with moisture in the upper atmosphere, chemical reactions occur and acid rain forms. In this way the pH of the rain is lowered. The lower the pH of the rainwater, the more acidic the rain.

The occurrence of acid rain is becoming more frequent in the world as the air becomes more polluted. Scientists believe that acid rain may affect plants and plant diseases. In some instances plants are stressed by acid rain and yield may be reduced. Acid rain may interact with plant pathogens and either reduce or increase the amount of damage from diseases.

The effects of acid rain are subtle and may not be immediately noticed. More research is needed to fully understand the possible long-term effects acid rain has on plant diseases and the production of food and fiber crops throughout the world.

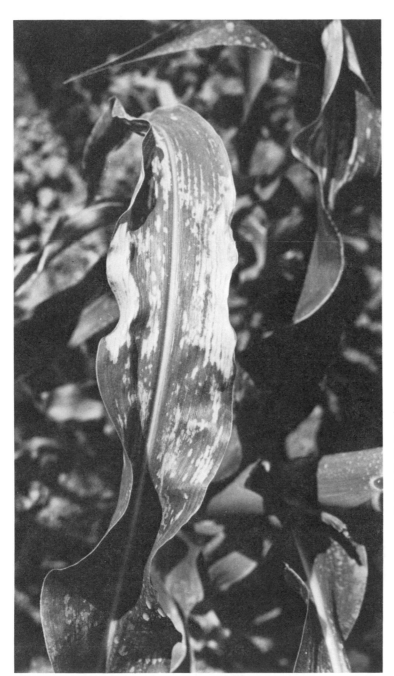

Fig. 20.7. Ozone and smog injury on corn. Air pollution frequently damages crops grown near big cities. The symptoms on corn can be confused with injury resulting from a postemergence application of a mixture of atrazine and nitrogen solution.
Courtesy N. C. Agric. Extension Service

Glossary

The terms included here usually appear in text discussions. Some specialized terms not listed here can be found in the Index.

A

Abscission.   The formation of a layer of cells that results in fruit, leaf, or stem drop from a plant.

Acre.   An area of land containing 43,560 ft^2 equal to 0.40 hectares.

Actinomycetes.   A group of microorganisms similar to bacteria that produce long filaments.

Adjuvant.   In relation to pesticides, any substance added to a formulation to improve its physical or chemical properties.

Adventitious roots.   Roots that appear in unusual places, e.g., on the stem.

Aeciospore.   A spore of the rust fungi produced in an aecium.

Aecium.   A fruiting body of the rust fungi in which aeciospores are produced.

Aerobic.   Relating to the growth of organisms in the presence of molecular oxygen.

Alternate host.   One of two different types of plants on which some pathogens must develop to complete their life cycle.

Amino acids.   The building blocks of proteins. There are 20 common amino acids; they are joined together in a strict order that determines the character of each protein.

Anaerobic.   Relating to biological processes that occur in the absence of molecular oxygen.

Annual.   A plant that completes its life cycle from seed to seed in 1 year.

Antagonism.   The phenomenon that results in a depression of activity of an organism when two or more occur in close proximity.

Anthracnose.   A leaf- or fruit-spot disease caused by fungi that produce spores in an acervulus.

Antibiotic.   A substance produced by one microorganism that inhibits or kills other microorganisms.

Antidote.   A substance used in treatment of pesticide poisoning.

Apothecium.   An open cup- or saucer-shaped structure in which some fungi produce spores (Ascomycetes).

Appressorium.   The swollen tip of a fungus hypha that attaches to and penetrates the host.

Aseptic.   Free of microorganisms.

Ascocarp. The fruiting body of some fungi (Ascomycetes).

Ascomycetes. A group of fungi that produce sexual spores called ascospores.

Ascospore. A sexual fungal spore produced by ascomycetes.

Ascus. A microscopic structure containing ascospores (usually eight).

Auxin. A plant growth-regulating substance that controls cell length.

B

Bacillus. A short rod-shaped bacterium.

Bacteriocide. A chemical that kills bacteria; usually an antibiotic.

Basidiomycetes. A group of fungi that produce sexual spores called basidiospores.

Basidiospore. A sexual fungal spore produced by basidiomycetes.

Basidium. A fungal structure on which basidiospores are produced.

Biennial. A plant that completes its life cycle in 2 years. The first year it produces leaves and store food; the second year it produces fruits and seeds.

Bioassay. Determination of the relative strength of a substance (such as a drug) by comparing its effect on a test organism with that of a standard preparation.

Biocide. A chemical that is toxic to many organisms, including both plants and animals.

Biological management. Techniques for managing pests by use of predators, parasites and disease-producing organisms.

Biosynthesis. The production of a chemical compound by a living organism.

Biotechnology. The use of biological systems and processes in industrial production; use of genetically engineered microorganisms.

Blight. A disease that causes rapid killing of many leaves, flowers, and stems.

Blotch. A disease with irregularly shaped spots on leaves, fruits, and stems.

Broad-spectrum pesticide. A chemical that controls a wide range of pests.

Budding. A method of vegetative propagation of plants by implantation of buds from the mother plant onto a rootstock.

C

Canker. A necrotic (dead) lesion on a stem or branch.

Capsid. The protein coat of viruses forming a shell around the nucleic acid.

Carbohydrate. A type of foodstuff composed of carbon, hydrogen, and oxygen; includes sugars, starches, and celluloses.

Carcinogen. A substance or agent capable of producing cancer.

Carrier. A liquid or solid material added to a chemical compound to prepare a proper formulation.

Cation. An ion having a positive charge.

Causal agent. The organism or abiotic factor that produces a given disease.

Cellulose. A polysaccharide composed of many glucose molecules and found in plant cell walls.

Centigrade (C).   A thermometer scale in which the freezing point of water is set at 0° and the boiling point at 100°. To change to degrees Fahrenheit, multiply degrees centigrade by nine-fifths and add 32. See also *Fahrenheit.*

Chemotherapy.   Control of a plant disease with chemicals that are translocated internally.

Chlamydospore.   An asexual fungal spore formed from a cell of a fungal hypha.

Chloroplast.   A structure in plant cells that contains chlorophyll and is the site of photosynthesis.

Chlorosis.   Yellowing of normally green plant tissue resulting from lack of chlorophyll.

Chromosomes.   The threadlike components of a cell that are composed of DNA and protein.

Circulative viruses.   Viruses that are acquired by a vector, accumulated internally, passed through vector tissues, and introduced into other plants.

Clone.   Many individual organisms produced asexually from one individual (e.g., rooted cuttings).

Coccus.   A spherical or round bacterial cell.

Compatible.   Two compounds that can be mixed without affecting each other's properties.

Concentration.   The amount of a chemical compound in a given volume or weight of diluent.

Conidiophore.   A specialized fungal hypha on which asexual spores (conidia) develop.

Conidium.   An asexual fungal spore formed on a conidiophore.

Contact pesticide.   A compound that kills by contact with plant tissue or pest rather than as a result of translocation.

Cotyledon.   The seed leaf. Monocotyledons have one seed leaf; dicotyledons have two.

Crop tolerance.   The ability of a crop to endure treatment with a chemical without adverse effect; also, the amount of a pesticide that is legally allowable on or in a crop at harvest time.

Cultivar.   Short for cultivated variety. See *Variety.*

Culture.   To grow microorganisms on a prepared medium.

Culture medium.   Prepared medium on which microorganisms are grown.

Cuticle.   A thin waxy layer on the outer wall of epidermal cells.

Cyst.   In fungi, an encysted zoospore; in nematodes, may contain eggs.

Cytoplasm.   Living substances in a cell except for the nucleus.

D

Damping-off.   Rapid dying of seedlings near the soil, resulting in the seedlings falling over; a disease caused by soilborne fungi.

Deciduous plant.   A perennial plant that loses its leaves during the winter.

Dermal.   Pertaining to the skin.

Dermal toxicity. Ability of a compound when absorbed through the skin of animals to produce symptoms of poisoning.

Detoxification. The inactivation or destruction of a toxin by alteration, binding, or breakdown of the toxic molecule.

Dieback. Death of shoots, branches, and roots starting at the tip.

Diluent. Any liquid or solid material used to dilute or carry an active ingredient.

Diploid. An individual with double the number of chromosomes (2N) per cell.

Disease. Abnormal development of a living organism that interferes with its normal structure and reduces its economic or aesthetic value.

Disease cycle. Stages involved in the development of a disease.

Disease tolerance. In plants, the ability of a host to survive and produce satisfactory yields at a level of infection that causes economic loss to other cultivars or types of the same host plant.

Disease triangle. A concept describing the simultaneous occurrence of a pathogen, a susceptible host, and a favorable environment such that a disease may develop.

Disinfectant. An agent that can be used to free plant tissue from infection.

Disinfestant. An agent that kills pathogens on a plant or plant part before infection occurs.

Dispersal. See *Dissemination.*

Dissemination. In relation to plant diseases, the transfer of inoculum to healthy plant tissue.

DNA (deoxyribonucleic acid). The genetic material found in all living organisms. Every inherited characteristic has its origin in the code of each individual's complement of DNA.

Dormant. Not actively growing (as seeds or other plant organs) due to internal causes.

Dose, Dosage. Quantity of a substance applied per unit of plant, soil, or other surfaces; also called *rate.*

Downy mildew. A plant disease, in which the fungus appears as a downy growth on the host surface.

Drift. In relation to pesticides, the movement of a portion of the airborne particles of a spray or dust away from the target area.

E

Economic poison. As defined under the Federal Insecticide, Fungicide and Rodenticide Act, an economic poison is "any substance or mixture of substances intended for preventing, destroying, repelling, or mitigating any insects, rodents, nematodes, fungi, or weeds, or any other form of life declared to be pests." As so defined, economic poisons are now generally known as *pesticides.*

Economic threshold. The pathogen density at or above which the value of crop losses—in the absence of management efforts—would exceed the cost of management practices (especially, the use of pesticides).

Ectoparasite. A parasite that feeds on the outside of a host.

Egg. In nematodes, the first stage of the life cycle.

Emulsion. A mixture in which one liquid is suspended as small globules in another liquid (e.g., oil in water).

Encyst. To form a cyst.

Endemic. Naturally occurs in geographical region.

Endodermis. A layer of cells that surrounds the vascular tissues of plant roots.

Endoparasite. A parasite that enters and feeds inside a host.

Enzyme. A protein that catalyzes a specific biochemical reaction.

Epidemic. A rapidly developing and widespread occurrence of a disease.

Epidermis. The outer layer of cells on all plant parts.

Epiphytotic. Relating to a widespread and destructive disease of plants (epidemic).

Eradication. Control of plant disease by eliminating the pathogen from a large area.

Etiolation. Elongation of stems caused by reduced light intensities.

Exclusion. Control of plant disease by excluding the pathogen or infected plant material from crop production areas.

Exudates. Liquid discharge from plant tissue.

F

Facultative parasite. An organism that is usually saprophytic but may become parasitic. See also *Obligate parasite.*

Facultative saprophyte. An organism that is usually parasitic but may live as a saprophyte.

Fahrenheit (F). A thermometer scale in which the freezing point of water is set at 32° and the boiling point is 212°. See also *Centigrade.*

Fermentation. Enzyme-catalyzed oxidation of organic substances in the absence of free oxygen.

Filamentous. Threadlike.

Fission. A form of asexual reproduction in which a bacterial cell splits into two cells.

Flagellum. A whiplike structure on bacterial cells or zoospores that functions in movement.

Flagging. The drooping of leaves and shoots before wilting of a plant.

Fleck. White to tan necrotic lesions up to a few millimeters in diameter, usually confined to the upper surface of leaves; a characteristic response of tobacco to ozone.

Formulation. In relation to pesticides, a specific mixture containing the active ingredient, the carrier, and other additives (adjuvants) required to make it ready for sale.

Fruiting body. A fungal structure in which spores are produced.

Fumigant. A toxic volatile substance that is used to manage various pests.

Fumigation. The application of a fumigant to an area.

Fungicide. A toxic substance to fungi; a type of pesticide.

Fungus. A microorganism composed of hyphae and lacking chlorophyll which is saprophytic and/or parasitic.

G

Gall. An overgrowth produced on a plant.

Gametangium. A male or female sexual organ in some fungi.

Gene. The hereditary unit; a segment of DNA coding for a specific protein.

Gene expression. The expression of the genetic material of an organism as specific traits.

Genetic code. The biochemical basis of heredity consisting of codons (base triplets along the DNA sequence) that determine the specific amino acid sequence in proteins and that are the same for all forms of life studied so far.

Genetic diversity. See *Genetic variability.*

Genetic engineering. Alteration of the hereditary apparatus of a living cell so that the cell can produce more or different chemicals, or perform completely new functions.

Genetic variability. The property or ability of a species to change its inherited characteristics from one generation to succeeding ones.

Genome. The basic chromosome set of an organism—the sum total of its genes.

Genotype. The genetic constitution of an individual or group.

Germ tube. The growth of hyphae from a germinating fungus spore.

Germicide. A substance that kills microorganisms.

Germination. The beginning of growth of a spore, or seed sprouting.

Germplasm. The total genetic variability available to an organism, represented by the pool of germ cells or seed.

H

Haploid. Having a single complete set of chromosomes.

Haustorium. A projection of specialized fungal hyphae into host cells which absorbs nutrients.

Hazard. In relation to pesticides, the probability that injury will result from use of a particular compound; the sum of the toxicity plus the exposure to a pesticide.

Hectare (ha). An area of land in the metric system equal to 2.47 acres.

Herbicide. A chemical used for killing or inhibiting plant growth; a weed or grass killer; a type of pesticide.

Hormone. A growth regulator (e.g., auxins and gibberellins in plants). See also *Plant growth regulator.*

Host. A living organism from which parasites or pathogens obtain nutrients.

Host range. The hosts that may be attacked by a pathogen.

Hyaline. Transparent; without color.

Hydathodes. Structures in which water moves from the interior to the surface of the leaf.

Hydrogen ion concentration. A measure of acidity. See *pH.*

Hypersensitivity. Plant tissues respond to certain pathogens by dying quickly and blocking their advance.

I

Immune. Cannot be infected by a given pathogen.

Incubation period. The time between penetration of a host by a pathogen and the appearance of symptoms.

Inert ingredient. Ingredient in a product which does not contribute to the activity of the active ingredient.

Infection. The early stage of disease caused by a pathogen.

Infectious disease. A disease that is caused by a pathogen that can be spread from a diseased to a healthy plant.

Infested. Containing or covered by great numbers of pathogens (insects, nematodes, bacteria, or fungi). Refers to the presence of pathogens in the environment, not to their infection of hosts.

Inoculation. The arrival of a pathogen on a host.

Inoculum. Part of a pathogen that comes in contact with a host and can cause disease.

Insecticide. A substance or mixture of substances that kills, repels, or mitigates insects; a type of pesticide.

Intercellular. Between cells.

Intracellular. Within cells.

In vitro. Outside the living body (e.g., a host) and in an artificial environment; in culture.

In vivo. In the living body of a plant or animal.

Ion. An electrically charged particle, atom, molecule, or radical in which the charge is due to the gain or loss of one or more electrons.

L

LD_{50}. A measure of toxicity; the dose of a compound that causes death in 50% of the test animals so treated. If a compound has an LD_{50} of 10 mg/kg, it is more toxic than one having an LD_{50} of 100 mg/kg.

Larva. The life stage of certain animals between the embryo and the adult (e.g., an immature nematode).

Leaching. Movement of a substance downward or out of the soil as the result of water movement.

Leaf spot. A lesion on a leaf.

Lesion. A localized area of diseased tissue.

M

Meristem. Formative plant tissue made of cells that can divide continually.

Metabolism. The processes by which organisms utilize nutrients to build structural components, and break down cellular material to obtain energy and simple substances for special functions.

Micrometer (μm). A unit of length equal to 1/1000 of a millimeter; also called *micron.*

Micron. See *Micrometer.*

Mildew. A disease of plants in which the fungus produces a whitish growth on the host surface.

Millimicron (mµ).   A unit of length equal to 1/1000 of a micrometer.

Millimeter (mm).   A unit of length equal to 1/10 of a centimeter (cm) or 0.039 of an inch.

Miscible.   Capable of being mixed and remaining mixed under normal conditions.

Mold.   Fungus growth on decaying matter or on the surface of plant tissues.

Monoculture.   The production of successive crops of the same plant species.

Mosaic.   Symptom of some viral diseases of plants characterized by normal and light green or yellowish patches on the foliage.

Mottle, mottling.   An irregular pattern of light and dark areas.

Mummy.   A dried, shriveled fruit.

Mycelium.   Mass of fungal hyphae that may be visible.

Mycoplasma.   A group of very small microorganisms, intermediate between viruses and bacteria.

Mycorrhiza.   A symbiotic relationship of a fungus with the roots of a plant.

N

Nanometer (nm).   A unit of length equal to one millionth of a millimeter.

Necrosis.   Localized death of tissue.

Necrotic.   Exhibiting varying degrees of dead areas or spots.

Nematicide.   A chemical compound that kills nematodes.

Nematode.   Any of a group of generally microscopic, eel-like worms that live as saprophytes in water or soil, or as parasites of plants and animals.

Noninfectious disease.   A disease that is caused by an abiotic agent that cannot be transmitted from one plant to another.

Nonionic.   Not having a positive or negative charge; electrically neutral.

Nonseptate.   Without crosswalls.

Nucleic acid.   Any of various acids composed of sugar or a derivation thereof, phosphoric acid and a base involved in transferring and expressing the genetic properties of organisms. See also *DNA; RNA.*

O

Obligate parasite.   An organism that can grow and multiply only on living organisms. See also *Facultative parasite.*

Organ.   A differentiated structure in an animal or plant made up of various cells and tissues and adapted for the performance of specific functions.

Organic compounds.   Chemical compounds that contain carbon.

Ozone.   A highly reactive form of oxygen that may injure plants.

P

Pandemic.   A widespread and destructive outbreak of disease simultaneously in several countries.

Parasite.   An organism that obtains food from another living organism (host).

Parts per billion (ppb).   A method of expressing the concentration of chemicals or other proportions. One pound is 1 ppb in 500,000 tons.

Parts per million (ppm).   A method of expressing the concentration of chemicals or other proportions. One inch is 1 ppm in 16 miles; 1 pound is 1 ppm in 500 tons.

Pasteurization.   The destruction, by heat or chemicals, of selected, harmful organisms (e.g., of plant pests) in soil or other propagating media.

Pathogen.   An agent that can cause disease.

Pathogenicity.   The capability of a pathogen to cause disease.

Penetration.   In relation to diseases, the initial invasion of a host by a pathogen.

Perennial.   A plant that lives for more than 2 years.

Pest.   Any life form that under some circumstances is detrimental to humans or to things they value (e.g., crops, stored food, domesticated animals).

Pesticide.   A chemical substance used to kill pests (weeds, insects, rats and mice, algae, nematodes, etc.). See also *Economic poison.*

pH.   A measure of acidity and alkalinity within the range of 0 to 14. The midpoint value (7) represents neutrality (neither acid nor alkaline), values below 7 have increasing acidity or hydrogen ion concentrations, and those above 7 have increasing alkalinity or lower hydrogen ion concentrations.

Photoperiod.   The optimum duration of light and darkness for the normal growth of a plant.

Photosynthesis.   The process by which plants combine carbon dioxide and water in the presence of light and chlorophyll to form carbohydrate.

Phycomycetes.   A group of fungi with few crosswalls in their hyphae.

Physiologic race.   A group of microorganisms similar to others in morphology but having distinct cultural, physiological, pathological, or other characteristics.

Phytopathogenic.   Capable of causing disease in plants.

Phytotoxic.   Poisonous, injurious, or otherwise harmful to plants.

Pith.   The parenchymatous tissue in the central area of a plant stem.

Plant-disease interaction.   The concurrent parasitism of a host by more than one pathogen in which the symptoms or other effects produced are of greater magnitude than the sum of the effects of each pathogen acting alone; an example of synergism.

Plant growth regulator.   A substance that affects plants or plant parts through physiological rather than physical action by accelerating or retarding growth, prolonging or breaking a dormant condition, promoting decay, or inducing other physiological changes.

Plasmid.   A spherical, self-replicating hereditary element that is not part of a chromosome. Because plasmids are generally small and relatively simple, they are used in recombinant DNA experiments as acceptors of foreign DNA.

Poison.   A substance that, when absorbed or taken orally can cause illness, death, or retardation of growth or shortening of life; also called *biocide.*

Polysaccharide.   A large organic molecule consisting of many units of a simple sugar.

Postemergence.   The period after the appearance of a specified weed or crop.

Ppb.   See *Parts per billion.*

Ppm. See *Parts per million.*

Predisposition. The tendency of non-genetic conditions acting before infection, to increase the susceptibility of a plant to disease.

Preemergence. The period before the emergence of a specified weed or crop.

Primary infection. The first infection of a plant by a pathogen.

Primary inoculum. Survival structures of a pathogen that cause primary infections.

Protectant. With pesticides, a substance that prevents infection by a pathogen.

Protective fungicide. A pesticide that protects an uninfected plant or seed by killing or inhibiting the development of fungus spores or mycelia on the surface.

Protein. A compound consisting of amino acids.

Protoplast. Part of a single cell inside the cell wall.

Protoplast fusion. The joining of two protoplasts or the joining of a protoplast with any of the components of another cell such that genetic transfer may occur.

Psi. Pounds per square inch.

Pustule. A small blisterlike structure in which some fungi produce spores.

R

Range. The geographic region throughout which a kind of organism lives. Not to be confused with *host range.*

Rate. The amount of pesticidal chemical (active ingredient) applied to a given surface area; also called *dose, dosage.*

Recombinant DNA. The hybrid DNA produced by joining pieces of DNA from different sources.

Relative humidity (RH). The ratio of the quantity of water vapor present in the atmosphere to the quantity that would saturate it at the same temperature.

Residue. In relation to pesticides, amount of chemical that remains on or in a harvested crop.

Resistance. The ability of an organism to overcome a pathogen or other damaging factor.

Resistant. Possessing qualities that hinder the development of a given pathogen or the effects of other damaging factors.

Resting spore. A fungal spore that is resistant to extremes in temperature and moisture and often germinates only after a period of time.

Restriction enzyme. An enzyme within a cell that recognizes and degrades DNA from foreign organisms, thereby preserving the genetic integrity of the cell. In recombinant DNA experiments, restriction enzymes are used as tiny biological scissors to cut up foreign DNA before it is recombined with a vector.

Rhizome. A modified underground rootlike stem that produces roots and leafy shoots (e.g., the white underground parts of johnsongrass).

Rhizosphere. The area near the surface of a living root.

Ringspot. A circular area of chlorosis which is characteristic of many virus diseases.

Rpm. Revolutions per minute.

RNA (ribonucleic acid). A nucleic acid involved in protein synthesis; also, the genetic material of many viruses.

Rogueing. The practice of removing diseased, inferior, or abnormal plants from a population.

Rosette. Short, bunchy type of plant growth.

Rot. The softening and disintegration of plant tissue by enzymes produced by fungal or bacterial infection.

Rust. A disease causing a "rusty" appearance on a plant.

S

Saprophyte. An organism that can obtain nutrients from dead organic material.

Scab. A crustlike diseased area on the surface of a plant organ.

Sclerotium. A compact mass of fungal hyphae, capable of surviving under unfavorable environmental conditions.

Scorch. "Burning" of leaf margins from disease or environmental conditions.

Secondary infection. Any infection from inoculum produced from primary or subsequent infections. See also *Primary infection.*

Selective pesticide. A chemical that is more toxic to some plant or animal species.

Septate. In fungi, hyphae with crosswalls.

Serological specificity. The pattern of reactions of a given antigen (e.g., protein from cell wall of a bacterium) and various antibodies and used in the detection and identification of antigenic substances.

Seta. A bristle-like protuberance from a spore or on some fungal fruiting bodies.

Sexual reproduction. Any type of reproduction that involves a union of male and female gametes.

Sign. Parts of a pathogen or its products seen on a diseased host plant.

Silvering, Silverleaf. A symptom of leaves or fleshy tissues characterized by greyish or shiny lesions; often induced by exposure to certain air pollutants.

Silviculture. The science of developing, caring for, or cultivating forests.

Soil sterilant. A chemical that, when present in the soil, prevents the growth of plants, microorganisms, etc.

Somatic cell. Cells other than sex or germ cells.

Sooty mold. A sooty fungal growth on foliage and fruit in the honeydew secreted by insects such as aphids.

Species. A class of related organisms having common characteristics and capable of interbreeding.

Sporangiophore. A specialized fungal hypha on which sporangia are produced.

Sporangium. Type of asexual fungal spores.

Spore.   In fungi, the reproductive unit consisting of one or more cells.

Sporodochium.   A fruiting structure of fungi made up of a cluster of conidiophores.

Sporulate.   Production of spores.

Stem pitting.   A symptom of some viral diseases with depressions on the stem.

Sterile fungi.   Fungi that are not known to produce spores.

Sterilization.   The elimination of pathogens (e.g., from soil or other rooting media) with heat or chemicals.

Stomata.   Minute, organized openings on the surface of leaves or stems through which gases pass.

Stylet.   A long, needle-like feeding structure of nematodes and some insects.

Succulent.   Tender, young, or watery plant tissues.

Surfactant.   A material that increases the emulsifying, dispersing, spreading, wetting, or other surface-modifying properties of a pesticide formulation.

Susceptibility.   The inability of a plant to overcome disease.

Susceptible.   Lacking the ability to resist the effect of a given pathogen, or other damaging factor.

Symbiosis.   A mutually beneficial relationship between two different organisms.

Symptom.   The visual reactions of an organism to a disease.

Systemic.   Spreading internally throughout the plant.

Systemic pesticide.   A pesticide that is translocated within the plant.

T

Teliospore.   Type of spore produced by rust and smut fungi.

Tip burn, tip necrosis.   Necrosis of apical tissues of leaves affecting only a small percentage of the entire leaf.

Tissue culture.   An *in vitro* method of propagating healthy cells from plant tissues.

Tolerance.   The ability of a plant to withstand the effects of a disease; also, the amount of toxic residue allowable on edible plant parts.

Tolerant.   Having the ability to withstand to a certain degree the effects of disease or other unfavorable circumstances.

Toxic.   Poisonous, injurious, or harmful to life.

Toxicant.   An agent capable of being toxic; a poison.

Toxicity.   The relative capacity of a compound to cause injury. See also LD_{50}.

Toxin.   A compound produced by an organism that is toxic.

Transpiration.   The loss of water vapor from plant surfaces.

Tumor.   An uncontrolled growth of tissue.

U

Uredospore.   A repeating spore produced by rust fungi.

V

Variability.   See *Genetic variability.*

Variety. A subdivision of a species where individuals within a species are distinct in form and function from other similar arrays of individuals.

Vascular. Relating to a plant tissue consisting of conductive tissue.

Vector. An agent able to transmit a pathogen.

Vein banding. Retention of bands of green tissue along the veins while the tissue between the veins has become chlorotic.

Vein clearing. Destruction of chlorophyll adjacent or in the vein tissue as a result of infection by a virus or other pathogen.

Vessel. In plants, a xylem element which function in conducting water and minerals.

Viricide. A substance capable of inactivating or suppressing the multiplication of a virus completely and permanently.

Virion. A virus particle.

Virulence. The degree of pathogenicity of a pathogen.

Virulent. Capable of causing a disease.

Virus. A obligate parasite composed of nucleic acid and protein.

Volatile. Able to vaporize (changes from a solid or liquid to a gas) at ordinary temperatures on exposure to air.

W

Wetting agent. A compound that reduces surface tension and causes a liquid to contact plant surfaces more thoroughly, usually a surfactant.

Wilt. Loss of rigidity of plant parts due to insufficient water in the plant.

Witches' broom. Broomlike growth from the dense clustering of branches.

X

Xylem. A complex tissue that typically constitutes the woody element in plants and functions in conduction, storage, and support.

Y

Yellows. A plant disease characterized by yellowing and stunting of plants.

Z

Zoospore. A spore capable of moving in water.

Suggested Readings

AGRIOS, G. N. 1978. Plant Pathology. 2nd ed. Academic Press, New York.

AINSWORTH, G. C. 1981. Introduction to the History of Plant Pathology. Cambridge Univ. Press, New York.

ALEXOPOULUS, C. J. and MIMS, C. W. 1979. Introductory Mycology. 3rd ed. Wiley-Interscience, Somerset, NJ.

ANON. 1984. Farm Chemicals Handbook. Meister, Willoughby, OH. (Published annually)

ANON. 1981. Impacts of Applied Genetics on Microorganisms, Plants, and Animals. U.S. Congress, Office of Technology Assessment, Washington, D.C.

ANON. 1984. North Carolina Agricultural Chemicals Manual. School of Agriculture and Life Sciences, N.C. State Univ., Raleigh. (Published annually)

BAKER, K. F. and COOK, R. J. 1974. Biological Control of Plant Pathogens. W. H. Freeman and Co., San Francisco.

BLANCHARD, R. O. and TATTAR, T. A. 1981. Field and Laboratory Guide to Tree Pathology. Academic Press, New York.

BREWER, J. W. and HARRISON, M. D. (eds.). 1973. Readings in Insect-Plant Disease Relationships. MSS Information Corp., New York.

CAREFOOT, G. L. and SPROTT, E. R. 1967. Famine on the Wind. Rand McNally, New York.

CHRISTENSEN, C. M. 1966. Molds and Man. Univ. of Minnesota Press, St. Paul.

COOK, R. J. and BAKER, K. F. 1983. The Nature and Practice of Biological Control of Plant Pathogens. Amer. Phytopath. Soc. St. Paul, MN.

COUCH, H. B. 1973. Diseases of Turfgrasses. R. E. Krieger Publ. Co., Huntington, NY.

COX, R. S. 1971. Private Practitioner in Agriculture. Solo Publications, Lake Worth, FL.

COX, R. S. 1982. The Agricultural Consultant. Publications Development of Texas, Crockett, TX.

CUMMINS, G. B. and HIRATSUKA, Y. 1983. Illustrated Genera of the Rust Fungi. Amer. Phytopath. Soc. St. Paul, MN.

DROPKIN, V. H. 1980. Introduction to Plant Nematology. Wiley–Interscience, New York.

FORSBERG, J. L. 1975. Diseases of Ornamental Plants. Special Publ. No. 3. Univ. of Illinois College of Agriculture, Urbana.

FRY, W. E. 1982. Principles of Plant Disease Management. Academic Press, New York.

GIBBS, A. and HARRISON, B. 1976. Plant Virology: The Principles. John Wiley and Sons, New York.

HEPTING, G. H. 1971. Diseases of Forest and Shade Trees in the United States. USDA, Forest Service. Agric. Handbook No. 386. U.S. Govt. Printing Office, Washington, D.C.

HORSFALL, J. G. and COWLING, E. B. 1977–1980. Plant Disease: An Advanced Treatise. Vols. I, II, III, IV, V. Academic Press, New York.

HUGHES, C. G. (ed.). 1964. Sugar-Cane Diseases of the World. Vols. 1 and 2. Elsevier, New York.

JOYNER, B. G. and LARSON, P. O. (eds.). 1980. Advances in Turfgrass Pathology. Harcourt Brace Jovanovich, Duluth, MN.

KAVALER, L. 1965. Mushrooms, Molds and Miracles. John Day Publ. Co., New York.

KENAGA, C. G. 1970. Principles of Phytopathology. Balt. Publications, Lafayette, IN.

KRANZ, J. (ed.). 1974. Epidemics of Plant Diseases. Springer-Verlag, New York.

LARGE, E. C. 1962. Advance of the Fungi. Dover Publ., New York.
LITTLEFIELD, L. (ed.). 1981. Biology of the Plant Rusts: An Introduction. Iowa State Univ. Press, Ames.
LUCAS, G. B. 1975. Diseases of Tobacco. 3rd ed. Biological Consulting Associates, Raleigh, NC.
MACE, M. E., BELL, A. A., and BECKMAN, C. H. 1981. Fungal Wilt Disease of Plants. Academic Press. New York.
MANION, P. D. 1981. Tree Disease Concepts. Prentice-Hall, Englewood Cliffs, NJ.
MARAMOROSCH, K. and HARRIS, H. F. (eds.). 1981. Plant Diseases and Vectors: Ecology and Epidemiology. Academic Press, New York.
MARAMOROSCH, K. and RAYCHAUDHURI, S. P. (eds.). 1981. Mycoplasma Diseases of Trees and Shrubs. Academic Press, New York.
MATTHEWS, R. E. F. 1976. Plant Virology. Academic Press, NY.
NELSON, P. E., TOUSSOUN, T. A., and COOK, R. J. (eds.) 1982. Fusarium: Diseases, Biology, and Taxonomy. The Pennsylvania State Univ. Press, University Park.
NELSON, R. R. (ed.). 1973. Breeding Plants for Disease Resistance: Concepts and Applications. The Pennsylvania State Univ. Press, University Park.
NYVALL, R. F. 1979. Field Crops Diseases Handbook. AVI Publishing Co., Westport, CT.
OU, S. H. 1972. Rice Diseases. Commonwealth Mycological Institute, Kew, England.
PANOPOULOS, N. J. (ed.). 1981. Genetic Engineering in the Plant Sciences. Praeger, New York.
PIRONE, P. P. 1978. Diseases and Pests of Ornamental Plants. 5th ed. John Wiley and Sons, New York.
PYENSON, L. 1981. Plant Health Handbook. A Guide to Better Gardening Indoors and Outdoors. AVI Publ. Co., Westport, CT.
ROBINSON, R. A. 1976. Plant Pathosystems. Springer-Verlag, New York.
SHARVELLE, E. G. 1979. Plant Disease Control. AVI Publ. Co., Westport, CT.
SHURTLEFF, M. C. 1966. How to Control Plant Diseases in Home and Garden. Iowa State Univ., Ames.
SMITH, K. M. 1977. Plant Viruses. 6th ed. John Wiley and Sons, New York.
STAKMAN, E. C. and HARRAR, J. G. 1957. Principles of Plant Pathology. Ronald Press, New York.
STEVENS, R. B. 1974. Plant Disease. Ronald Press, New York.
STREETS, R. B. 1969. Diagnosis of Plant Diseases. Agric. Ext. Serv., Univ. of Arizona, Tucson.
STROBEL, G. A. and MATHRE, D. E. 1970. Outlines of Plant Pathology. Van Nostrand, New York.
TARR, S. A. J. 1972. Principles of Plant Pathology. Winchester Press, New York.
TAYLOR, A. L. and SASSER, J. N. 1978. Biology, Identification, and Control of Root-Knot Nematodes (Meloidogyne species). A Coop. Pub. of Dept. of Plant Pathology, N.C. State Univ. and U.S. Agency for International Development. NCSU Graphics, Raleigh, NC.
VANDERPLANK, J. E. 1975. Principles of Plant Infection. Academic Press, New York.
VANDERPLANK, J. E. 1978. Genetic and Molecular Basis of Plant Pathogenesis. Springer-Verlag, New York.
WALKER, J. C. 1969. Plant Pathology. 3rd ed. McGraw-Hill, New York.
WEBSTER, J. M. (ed.). 1972. Economic Nematology. Academic Press, New York.
WELLMAN, F. L. 1971. Plant Diseases—An Introduction to the Layman. Natural History Press, Garden City, NY.
WELLMAN, F. L. 1972. Tropical American Plant Diseases. Scarecrow Press, Metuchen, NJ.
WESTCOTT, C. 1971. Plant Disease Handbook. 3rd ed. Van Nostrand, New York.
WHEELER, B. E. J. 1969. Introduction to Plant Diseases. John Wiley and Sons, New York.
ZADOKS, J. C. and SCHEIN, R. D. 1979. Epidemiology and Plant Disease Management. Oxford University Press, New York.
ZUCKERMAN, B. M. and ROHDE, R. A. (eds.). 1981. Plant Parasitic Nematodes. Vol. III. Academic Press, New York.

The following publications of the American Phytopathological Society (3340 Pilot Knob Rd., St. Paul, MN 55141) also provide valuable information on plant diseases (publication dates in parentheses):

Compendium of Alfalfa Diseases	(1979)
Compendium of Barley Diseases	(1982)
Compendium of Corn Diseases	(1980)
Compendium of Cotton Diseases	(1981)
Compendium of Elm Diseases	(1981)
Compendium of Pea Diseases	(1984)
Compendium of Peanut Diseases	(1983)
Compendium of Potato Diseases	(1981)
Compendium of Rose Diseases	(1983)
Compendium of Soybean Diseases	(1982)
Compendium of Strawberry Diseases	(1984)
Compendium of Turfgrass Diseases	(1983)
Compendium of Wheat Diseases	(1977)

Index _____

Colchicine, 108
Colletotrichum spp., 43, 204
Common names
  of diseases, 11–12
  of pesticides, 60
Conidia, 143, 182, 186–190, 193–196, 199, 206
Conidiophore, 190, 193, 199
Control of plant diseases (*see* Management of plant diseases)
Controlled droplet application (CDA), 75
Copper naphthenate, 226
Coremia, 216
Corn
  Anthracnose, 202
  Fusarium rot, 161
  Helminthosporium spp., 196–198
  lesion nematode, 134
  maize chlorotic dwarf, 254, 266–268
  maize dwarf mosaic, 254, 266–268
  Rhizopus soft rot, 209
  smut, 177–178
  southern leaf blight, 196–199
  stem and bulb nematode, 128, 136
  stubby root nematode, 128, 136
Corn smut, 177–178
Corn stunt, 267
Cortex, 135
Corynebacterium michiganese, 233, 242
Covered smut, 178–179
Creosote, 226
Cronartium fusiforme, 212, 220–221
Crop Improvement Association, 54
Cropping sequence, 42
Cropping systems, 41–45
Crown gall, 233, 244–246
Cucumber
  Alternaria leaf spot, 200
  Anthracnose, 203
  bacterial soft rot, 239
  bacterial wilt, 233, 240–242
  belly rot, 239
  cucumber mosaic, 254, 263
  cucumber pale fruit, 268
  Rhizoctonia solani, 156
Cucumber mosaic, 254, 263–265
Cucumber mosaic virus (CMV), 34, 254, 264–265
Cucurbits
  Anthracnose, 202–203
  bacterial wilt of cucumber, 233, 240–242
  powdery mildew, 186–189
  Rhizopus soft rot, 209
Curative, 59
Cuscuta spp., 269, 273
Cylindrocladium crotalariae, 144, 167–168
Cytokinins, 108

D

Damping-off diseases, 151–154
Dagger nematode, 138
DeBary, H. A., 15
Debilitating disease, 4
Deficiencies, nutrient, 275–276
Density of liquids, 91
Devastating diseases, 4
Diaphragm, pump, 82
Diploid cells, 108
  growth hormones, 108
  regeneration of haploid cells, 108
Disease complexes, 11, 123
Disease cycle, 19–22
Disease resistance, 49–53
  developing resistant cultivars, 52–53
  genetic uniformity, 50
  kinds of, 51–52
Ditylenchus dipsaci (*see* Stem and bulb nematode)
DNA, 101, 110–114, 250, 252
  bacterial vectors, 112–113
  transplanting from one cell to another, 110–114
  viral vectors, 114
Dodder, 269, 273–274
Downy mildews, 180–186
  of cucurbits, 185–186
  of grape, 184–185
  of potato, 180–183
Dutch elm disease, 212, 215–216
Dwarf mistletoe, 269, 271–273

E

Early blight, 200
Economic thresholds, 30
Ecosystem, 97
Ectomycorrhizae, 222
Ectoparasites, 9
Ectoparasitic nematodes, 9, 120, 137–139
  dagger nematode, 138
  lance nematode, 138
  ring nematode, 138
  sting nematode, 138
  stubby root nematode, 137–140
  stunt nematode, 138
Edaphology (soils), 3
Electrostatic, 75
Emulsion, 64
Endomycorrhizae, 222
Endoparasites, 9
Endoparasitic nematodes, 9, 120, 137
  lesion nematode, 134–136
  root knot nematode, 128–134
Endothia parasitica, 212, 214

powdery mildew, 186
    stubby root nematode, 128, 139
Greenhouse management, 45–46
    pesticide application, 88
Ground speed, 91
Gymnosporangium juniperus–virginianum, 170, 175

H

Haploid cells, 108
Haustoria, 182, 188
Helminthosporium spp., 50, 196–200
    maydis, 108, 197–198
    oryzae, 199
Heterobasidion annosum, 212, 219
Heterozygous, 103
High volatile fumigants, 70
Hollow stalk, 239
Horizontal resistance, 51
Hydathodes, 21
Hydraulic agitator, 83
Hyphae, 9, 141
Hyphal fusion, 174
Hypovirulent, 214

I

Immune, 51
Impregnating, 64
Incubation period, 22
Indicator plants, 254
Infection court, 47, 51
Infection cushion, 159
Infection pegs, 21
Infectious plant diseases, 8–11
Inoculation, 21
Integrated pest management, 26–31, 183, 196
    economic threshold, 30
    evolution, 26–28
    IPM concept, 29–30
    management principles, 28
Inversion, 99
IPM (*see* Integrated pest management)
Irish famine, 15, 180

J

Jet agitator, 83
Juveniles, 9

K

Koch's Postulates, 6
Knots (*see* Galls)

L

Lance nematode, 138
Larvae, 9, 120, 132, 136, 137
    Meloidogyne spp., 131–132
    Pratylenchus spp., 136
    Ditylenchus spp., 137
Late blight, 15, 180–183, 200
Leaf blight, 156
Leafy mistletoe, 269, 271–273
Lesion nematode, 134–136
    disease cycle, 136
    Pratylenchus brachyurus, 136
    symptoms, 134–136
Lespedeza
    dodder, 274
    Rhizoctonia solani, 156
Limiting diseases, 4
Little leaf of pine, 212, 222–223
Local infection, 177
Loose smut, 179–180
Low-volatile fumigants, 68

M

Maize chlorotic dwarf, 254, 266–268
Maize chlorotic dwarf virus (MCDV), 254, 267–268
Maize dwarf mosaic, 254, 266–268
Maize dwarf mosaic virus (MDMV), 254, 266–268
Male sterility, 110, 196–199
    southern leaf blight of corn, 196–199
    Texas cytoplasm, 197
Management of plant diseases, 32–57
    avoidance of pathogen, 33
    certified seed, 54–55
    chemicals, 58–100
    cropping systems, 41–45, 124–125
    disease resistant cultivars, 49–53
    environment, adjustment of, 47
    eradication, 33–35
    evasion, 32–33
    exclusion, 32–33
    greenhouse management, 45–46
    inoculum reduction, 35–47
    insect management, 47
    nematodes, 124–127, 137
    physical methods, 48–49
    rotations, 41–45
    sanitation, 35–41
    seed storage, 55–56
    weed management, 46–47
Macroconidia, 161
Major elements, 275
Mechanical agitator, 83